THE INFINITE IN MATHEMATICS

VIENNA CIRCLE COLLECTION

VOLUME 9

EDITOR: BRIAN McGUINNESS

FELIX KAUFMANN (1895–1949)

FELIX KAUFMANN

THE INFINITE IN MATHEMATICS

Logico-mathematical writings

Edited by

BRIAN McGUINNESS

with an introduction by

ERNEST NAGEL

translated from the German by

PAUL FOULKES

D. REIDEL PUBLISHING COMPANY

DORDRECHT : HOLLAND / BOSTON : U.S.A.
LONDON : ENGLAND

Library of Congress Cataloging in Publication Data

CIP

Kaufmann, Felix, 1895–1949.
The infinite in mathematics.

(Vienna circle collection; 9)
'Bibliography of published works by Felix Kaufmann': p.
Bibliography: p.
Includes index.
1. Infinite. 2. Mathematics—Philosophy. I. McGuinness, Brian.
II. Title. III. Series.
QA9.K313 510'.1 78-9123
ISBN-13: 978-90-277-0848-9 e-ISBN-13: 978-94-009-9795-0
DOI: 10.1007/978-94-009-9795-0

The main essay in this collection,
DAS UNENDLICHE IN DER MATHEMATIK UND
SEINE AUSSCHALTUNG
was first published by Franz Deuticke, Vienna, 1930

Published by D. Reidel Publishing Company,
P.O. Box 17, Dordrecht, Holland

Sold and distributed in the U.S.A., Canada, and Mexico
by D. Reidel Publishing Company, Inc.
Lincoln Building, 160 Old Derby Street, Hingham,
Mass. 02043, U.S.A.

Softcover reprint of the hardcover 1st edition 1978

TABLE OF CONTENTS

INTRODUCTION

The main item in the present volume was published in 1930 under the title *Das Unendliche in der Mathematik und seine Ausschaltung*. It was at that time the fullest systematic account from the standpoint of Husserl's phenomenology of what is known as 'finitism' (also as 'intuitionism' and 'constructivism') in mathematics. Since then, important changes have been required in philosophies of mathematics, in part because of Kurt Gödel's epoch-making paper of 1931 which established the essential incompleteness of arithmetic. In the light of that finding, a number of the claims made in the book (and in the accompanying articles) are demonstrably mistaken. Nevertheless, as a whole it retains much of its original interest and value. It presents the issues in the foundations of mathematics that were under debate when it was written (and in some cases still are); and it offers one alternative to the currently dominant set-theoretical definitions of the cardinal numbers and other arithmetical concepts.

While still a student at the University of Vienna, Felix Kaufmann was greatly impressed by the early philosophical writings (especially by the *Logische Untersuchungen*) of Edmund Husserl. He was never an uncritical disciple of Husserl, and he integrated into his mature philosophy ideas from a wide assortment of intellectual sources. But he thought of himself as a phenomenologist, and made frequent use in all his major publications of many of Husserl's logical and epistemological theses. He had been a student of the legal philosopher Hans Kelsen at the University, received the doctorate in law in 1920 and the doctorate in philosophy two years later, and on Kelsen's recommendation was appointed in the latter year to the unsalaried post of *Privatdozent* of the philosophy of law in the juridical faculty of that institution.

To earn a living, Kaufmann sought employment in business, and eventually became the manager of the Austrian branch of the Anglo-Iranian Oil Company. Nevertheless, by the time he came to write the present book he had published three others on the philosophy of law. He undertook in them to recast Kelsen's 'pure theory of law' by substituting for its neo-Kantian assumptions, which Kaufmann found unsatisfactory, a phenomenological epistemology. He found time to be a fairly regular attendant at the meetings of the *Wiener Kreis* (or the *Vienna Circle*), a discussion group of philosophers and scientists organized by Moritz Schlick after he joined the faculty of the University as professor of philosophy, which eventually acquired international repute as the progenitor of logical positivism (or logical empiricism, as the movement was also called). However, although Kaufmann was in wholehearted sympathy with many of its attitudes – especially with its stress on clarity and logical rigor in the conduct of philosophical inquiry – he objected to being counted as a logical positivist, and he saw himself as constituting the loyal opposition to the atomistic empiricism of the *Kreis*. Kaufmann's indebtedness to Husserl continued to be exhibited in his *Methodenlehre der Sozialwissenschaften* published in 1936, as well as in its completely rewritten version published in 1944 with the title *Methodology of the Social Sciences*. The latter work, written in English after he left Hitler's Vienna to join the Graduate Faculty of the New School for Social Research in New York City, also reveals the influence on his thought of John Dewey's logical theory when he became familiar with it in the United States. Kaufmann died unexpectedly in 1949 at the age of 54 years.

This summary account of Kaufmann's career makes evident the unusually broad scope of his active scholarly interests. This breadth flowed directly from his conception of the general task of philosophy. As he saw it, that task is to provide an ongoing critique of knowledge by articulating the logical conditions discourse must satisfy to be meaningful, making explicit the rules governing the acceptance and rejection of beliefs, and thereby producing the intellectual tools for clarifying and evaluating unsettled issues in various branches of inquiry. In consequence, the central objective of his wide-ranging studies was to make manifest the principles men employ when they succeed in making their experience intelligible, and to assess in the light of those principles disputed claims to knowledge in a number of special disciplines.

Kaufmann's pursuit of these objectives was controlled by a variety of special assumptions, commonly though not exclusively made by phenomenologists. The most inclusive of these, the Principle of Phenomenological Accessibility, asserts that whatever has a locus in any realm of being is 'accessible to cognition', so that there is nothing inherently unknowable or incapable of precise analysis. Although it is not entirely clear what it means to be 'accessible to cognition', Kaufmann used this Principle to show that concepts apparently referring to things transcending all possible experience either have no function in the acquisition and edifice of knowledge (and are therefore eliminable), or have meanings specifiable in terms of experimentially identifiable procedures. Kaufmann's approach to the task of clarifying ideas has much in common with the 'operationalism' of P. W. Bridgman and other instrumentalist thinkers, though he rejected the sensationalistic epistemology to which Bridgman subscribed.

It is this 'procedural approach' to the analysis of concepts that characterizes Kaufmann's philosophy of mathematics and the discussions in the present work. The reform movement in the 19th century known as 'the arithmetization of mathematics' sought to remove serious obscurities and confusions in various branches of mathematical analysis by redefining all the concepts used in them (such as the notions of imaginary and irrational number, continuity, or the derivative of a function) in terms of the familiar arithmetical operations upon integers; and in consequence, the assumption that dubious 'entities' such as infinitesimals are needed for differentiating or integrating functions was shown to be unnecessary. However, with the development of set-theory during the second half of the century – the theory came to be regarded eventually as the foundation for the rest of mathematics – 'objects' of a new sort were introduced into the subject (such as nondenumerably infinite classes and transfinite numbers), which many outstanding mathematicians, among others E. Borel, L. E. J. Brouwer, and H. Weyl, believed were as questionable as was the assumption of infinitesimals. (But the notion of infinitesimals has been placed on secure foundations during the past twenty five years, so that the 19th-century objections to infinitesimals cannot be validly raised against the revised notion. Infinitesimals have become respectable.) Kaufmann shared this belief, and the present book is his attempt to show that contrary to appearances mathematics nowhere requires the notion of an 'actual infinite', and that the standard arguments for the 'existence' of various orders of infinity are fallacious.

Kaufmann subscribed to the familiar Leibnizian distinction between necessary truths of reason and contingent truths of empirical fact. The former are certifiable by examining the meanings (or connotations) of the terms contained in them; the latter are based on the outcome of observation or experiment, and are in principle always corrigible. In consonance with this fundamental dichotomy, he followed Husserl in distinguishing between two sorts of universal statements, called 'specific' and 'individual'. A *specific universal* (such as the statement 'All prime numbers greater than two are odd') is said to assert that a specified relation holds between *concepts*, so that deductive logic allegedly suffices to establish the truth or falsity of the statement. On the other hand, an *individual universal* (such as the statement 'All the animals exhibited in the Milwaukee Zoo during the current year weigh more than two pounds') is characterized as being, in effect, the conjunction of a finite number of singular statements about certain individuals in specified spatio-temporal region, so that the truth-value of the universal can be determined by examining those individuals seriatim. It is therefore alleged that it is a blunder to suppose that the sense of the particle 'all' in specific universals is the same as the sense of 'all' in individual universals, or that specific and individual universals can be construed in identical ways. In Kaufmann's judgment, however, it is just such a conflation of two radically different meanings of 'universal statement' which is at the bottom of the allegedly mistaken beliefs that cardinal numbers are logically prior to the ordinals, and that there are infinite classes whose members are non-denumerable.

Kaufmann's book must be consulted for the details of his argument for these conclusions, and only so much of its salient features will be mentioned as is needed to make intelligible some comments on several of his major claims. In contrast to the Frege-Russell definition of the cardinal numbers as classes (or sets) of similar classes, Kaufmann defined them as invariants of counting processes – that is, in whatever order the members of a collection are matched with the members of some serially ordered set of standard items, such as the numerals, the last member of the collection to be matched will always correspond to the same numeral. Accordingly, though the cardinal number of a collection is independent of *any one* order in which its members are counted, it is not independent of *all* such orders, so that ordinal numbers are logically prior to the cardinals. On the other hand, the ordinal numbers themselves are defined, in a manner substantially in agreement with the Peano axioms for arithmetic, as the formal structures embedded in counting processes that have no fixed termination. In consequence, the phrase 'the infinite series of the integers' must not be taken to denote an 'actual infinite totality', as if statements about all the integers were individual universals. On the contrary, the phrase is said to be a term that enters into specific universals, and to signify the formal serial structures that are distinctive of processes of counting. In the case of a finite collection of items, it makes good sense to talk about the class of all its sub-classes, for not only can each of these sub-classes be 'constructively' defined (that is, a rule or 'law' can be stated for obtaining each of them), but the set (or class) of *all* these sub-classes can also be constructively defined. However, on this construal of the term 'set' or 'class' (according to which, quoting Kaufmann, 'an infinite set is nothing but a law'), the expression 'the class of all sub-classes of the class of all integers' is meaningless. For the so-called 'class' mentioned in the expression is not, and cannot be, constructively defined, so that it is also meaningless to assert that 'the members of this class' are non-denumerable. Moreover, Kaufmann maintained that his definition of the integers established not only the *consistency* of Peano's axioms (that is, that there is no arithmetical statement such that both the statement and its denial are derivable from the axioms), but also their *completeness* (that is, that there is no arithmetical statement such that neither the statement nor its negation is derivable from the axioms).

Kaufmann's rejection of Cantor's 'diagonal proof' that the 'totality' of the real numbers (or the continuum) is non-denumerable is based on the same considerations that led him to reject as meaningless the expression 'the class of all sub-classes of the set of integers'. For the diagonal proof proceeds on the assumption that it makes sense to talk about the totality of all the reals, an assumption Kaufmann denied on the ground that the alleged totality is not constructively definable, and that the assumption confounds the sense of 'all' in specific universals with the sense of the particle in individual ones. According to him, what the diagonal proof does establish is that for any given denumerable sequence (i.e., for any sequence specified by some determinate rule for constructing its members) of denumerable sequences of integers (which are also specified by some constructive rule), another sequence of integers can be defined (i.e., another rule of construction can be formulated) which is not included in the initial rules of construction. But it does not follow from this conclusion, so he maintained, that the real numbers (or rules for constructing sequences of integers) form a totality and that they are non-denumerable.

Although Kaufmann was not alone in defining cardinal numbers in terms of the ordinals or in rejecting as absurd the notion of non-denumerable classes, the reasons he gave for these views were in considerable measure his own. Moreover, unlike many who arrived at similar conclusions (notably the mathematician Brouwer), he believed that the formal structures investigated by mathematicians are discoveries rather than human creations, and that the constructive intuitionism to which he subscribed does not require the rejection as false of any principles of classical logic (such as the principle of excluded middle).

Kaufmann was unquestionably correct in holding that the cardinal numbers *can* be defined in terms of the ordinals. The mode of defining the cardinals he proposed has some clear advantages over the alternative Frege-Russell procedure – for example, his method makes more evident than does the set-theoretical definition the function of cardinal numbers in normal every-day and scientific practice. However, he did not recommend his way of defining the cardinals for such pragmatic reasons. He did so because he believed that the set-theoretical definition is fundamentally unsound and involves a serious blunder. But it is not obvious that Kaufmann succeeded in showing this to be the case, and it is therefore appropriate to ask whether he did in fact accomplish this.

The premises on which he based the conclusion that the set-theoretical definition is unsound include the assumption – let us grant it without discussion for the sake of the argument – that the definition involves the interpretation of specific universals as if they were individual universals. Kaufmann's case against the set-theoretical definition then depends on whether it is a hopeless error, as he believed, to suppose that terms occurring in specific universals have *extensions* and that the extensions are *classes* of items. In agreement with a long tradition in philosophy, the distinction between specific and individual universals must be admitted to be well-founded; and it is at least plausible if not true that terms occurring in specific universals are in general not associated with any extensions at all, or that if the terms do have extensions the extensions are not classes. But it by no means *follows* from these premises that it is an error to modify common usage by *stipulating* that in certain contexts classes are to be the extensions of such terms. If it is an error nonetheless, Kaufmann has not shown that it is. On the other hand, if it is not an error, the set-theoretical definition of cardinals is a viable alternative to his definition of them in terms of the ordinals. In that case, the question which mode of definition is the preferred one can then have no *a priori* answer, and can be decided only after ascertaining the relative merits of the two modes of definition in making it possible to attain specific objectives. It is conceivable, for example, that the ordinal definition is better suited for performing one task (e.g., clarifying the nature of counting and the logic of measurement), while the set-theoretical definition is more useful in undertaking another (e.g., providing a set-theoretical foundation for a comprehensive systematization of the various branches of mathematics).

Kaufmann's argument for his contention that the idea of non-denumerable infinities is absurd, is also inconclusive for the reasons just stated, so that nothing further need be said about it. Moreover, although he used the important notion of *constructive* definitions and proofs, he used it in an informal, intuitive manner, without stating precisely just what is the distinction between constructive and non-constructive definitions and proofs in mathematics. Indeed, the distinction was not clearly formulated until the theory of recursive functions was developed after the publication of this book. Kaufmann also believed that the so-called 'second order' (or 'higher') logical calculus – which deals with statements ascribing properties (or attributes) to properties – is not needed in general, and in particular not in mathematics. For example, the statement 'The relation of being greater in magnitude is asymmetrical' is a second-order statement, since it ascribes the property of being asymmetrical to the relational property of being greater in magnitude. But this second-order statement is eliminable, for its content is fully rendered by the first-order statement 'If one "object" (e.g., a number) is greater in magnitude than a second, then the second is not greater in magnitude than the first'. However, although many second-order attributions are eliminable because they are logically equivalent in content to first-order statements, this cannot be done always. This becomes evident in defining the notion of one number being *a successor* of another in a sequence of numbers generated by the relation of one number being *the immediate successor* of another. The required definition can be stated as follows: 'y is a successor of x' if and only if 'There is a class of numbers C of which y is a member but x is not, and every number z belonging to C is either the immediate successor of x or is the immediate successor of some number in C'. It is clear that the definition makes mention of a certain class C which is described in terms of its members and has the described property ascribed to it; and since the definition contains the class term C existentially quantified, the term is not eliminable.

Some of these critical comments on Kaufmann's claims are doubtless debatable. However, it is no longer a matter for serious debate whether his account of the structure of counting also established, as he believed, the consistency as well as the completeness of Peano's axioms. For the untenability of this belief became evident with the appearance in 1931 of the Gödel paper to which reference has already been made. Kaufmann's book was published a year earlier, and his claim concerning the consistency and completeness of arithmetic was not wholly unwarranted at the time it was made. Although a number of the views presented in this book must be corrected in the light of later developments in the subject, the book was never revised; nor did Kaufmann leave any indications of what changes in his philosophy of mathematics he thought were made necessary by Gödel's discoveries. But despite these limitations, his book remains an enlightening and stimulating contribution to a fundamental branch of philosophical inquiry.

ERNEST NAGEL

Columbia University

EDITOR'S NOTE

Felix Kaufmann (1895–1949) represented, in the way described in Ernest Nagel's introduction, the intersection of the Vienna Circle and the phenomenological movement. His thinking may fairly be said to combine the merits of the two schools. We publish here his main writings in logic and mathematics. Chief of these is the work on the infinite (its author's favourite book): grateful acknowledgement for permission to publish a translation of this must go to Franz Deuticke of Vienna, the house (happily still flourishing) which first published it in 1930. The book was reprinted in German by the Wissenschaftliche Buchgesellschaft of Darmstadt in 1968. There follows an article from *Erkenntnis* **2** (1931), for permission to publish a translation of which we are indebted to the house of Felix Meiner. An early version of this article found among Kaufmann's papers is marked 'Schlick Kreis 13.XI.1930' and was no doubt delivered as a lecture on that day. Finally we have included an unpublished paper of about 1931. This was kindly supplied by the Centre for Advanced Research in Phenomenology at Wilfrid Laurier University, Waterloo, Ontario, where Professor José Huertas-Jourda and Dr. Harry P. Reeder were in every way most helpful. Dr. Reeder has put us further in his debt by a bibliography of Kaufmann's publications, designed for the present volume.

As the editor responsible in this case I am particularly indebted to Dr. Else Kaufmann, the author's widow, who greeted visits and enquiries with encouragement rather than patience. She and their son, Mr. George Kaufmann, have helped to preserve Kaufmann's work and have agreed most readily and on most generous terms to its publication when that was urged upon them. Happy in his heirs, in name, in nature, and only not in length of life, Felix Kaufmann seems to us to merit study for the variety of his gifts and for the particular turn he gave to the ideas of the Vienna Circle. Two further volumes of his writings are planned.

B. McGUINNESS

THE INFINITE IN MATHEMATICS
AND ITS ELIMINATION

An Enquiry into the Foundations of Mathematics

1

PREFACE

The present work deals with the problems of the foundations of mathematics, which are largely connected with the concept of the infinite, and attempts to reach a clear decision on the most important controversial questions. Its subject is therefore philosophy of mathematics.

The word 'philosophy' is here understood not in the sense of speculative construction, unfortunately all too common, but rather means clarification of thought through reflection. Therein lies the central task of a philosophy to be distinguished from the positive sciences: this view is steadily gaining ground amongst the leading thinkers of our time, in spite of all metaphysical 'adventures of reason', and will consequently put an end to philosophical 'orientations', whose differing terminologies often mask extensive agreement in fact.

The task of a philosophy of mathematics, then, consists in bringing out with full clarity, the sense of mathematical propositions, concepts and method. That the posing of this problem is not only in principle justified, but has been topical for mathematicians these thirty years past (since the appearance of the antinomies), none who are familiar with foundational questions in mathematics will deny. Amongst the analyses thus given in outline, special emphasis belongs to critical investigations of method pertaining to the discovery and examination of presuppositions used in mathematical procedures. That such examinations can quite generally lead to significant results, has been shown especially by examples from modern physics; witness for instance the transformation of the concept of simultaneity of spatially distant events within the framework of relativity theory, and the revision of hitherto prevailing views of causality in connection with quantum theory.

As we shall see, a critical analysis of mathematical method results in the elimination of those assumptions that lead to the introduction of the actual infinite, and in particular the non-denumerable infinite, into mathematics; but this elimination in no way weakens the structure of classical mathematics, as is often alleged.

As is clearly evident from our initial remarks, the following investigations will not rest on any 'philosophical position', and therefore must not introduce any external presuppositions into the states of affairs to be analysed: these latter alone will have to guide us.

What first made me appreciate this objective attitude were the philosophical works of Edmund Husserl; for this, though not for this alone, I shall forever owe him my deep gratitude.

It is some thirteen years since the study of the logical and set theoretical antinomies first led me to the problems discussed in this essay and I have constantly returned to them. For once you have properly recognized and grasped them, and seen what essential part of logical and epistemological problems they contain, in pure culture as it were, you cannot break loose from them again.

Nevertheless, this essay would hardly have come into being, had I not constantly received important hints through attending the discussions of a circle of philosophers and mathematicians in which, under the direction of M. Schlick and H. Hahn, problems of the foundations of mathematics have been considered these last six years.

In general, time and again over recent years, I have sought and found occasion to discuss the relevant problems, in order to avoid or remove defects of substance or exposition in their treatment. For help in these endeavours my sincere thanks go to Oskar Becker (Freiburg i. Br.), Adolf Fraenkel (Kiel), Moritz Geiger (Göttingen), Hans Hahn (Vienna), Carl Gustav Hempel (Berlin), Karl Menger (Vienna), Friedrich Waismann (Vienna); and most particularly to Heinrich Behmann (Halle a. d. Saale) and Rudolf Carnap (Vienna).

Those writings, which my investigations rely on or criticize, will be adequately noticed in the text and in the notes, so that I need not mention them specially here. At this point I wish merely to point to a similarity of view concerning the question of the non-denumerable infinite, between myself and the great French writers on the theory of functions Borel, Baire and Lebesgue, because this does perhaps not come out clearly enough in the essay itself.

In the bibliography at the end of this book I mention only works that have been referred to in the notes; for an almost complete list of all the more important publications (up till mid-1928), see A. Fraenkel, *Einleitung in die Mengenlehre*, 3rd edition.

The present work is in the first place addressed to readers who are already familiar with the relevant problems, but with the exception of a few passages in text and footnotes it does not presuppose any special prior knowledge of mathematics and symbolic logic. It is therefore accessible to enquirers and students who are mainly interested in general problems of epistemology and have only a slight grasp of mathematical and logical techniques.

Vienna, December, 1929

ANALYTIC TABLE OF CONTENTS

arising in the Russell–Whitehead construction of mathematics – Irrational roots of equations – Preserving the knowledge attained in mathematical analysis – Irrational numbers and geometrical 'intuition'.

V. SET THEORY

Recapitulation of results bearing on the investigation that follows – Finite and infinite sets – The problem area of Cantor's set theory – Account of the calculus of powers – Subset and equivalence – Transfinite cardinals – Representation of infinite sets on their proper subsets – Dedekind's definition of an infinite set – Denumerable sets – The diagonal procedure – Critical analysis of this – Interpretations going beyond the mathematical state of affairs in set theory and consequences thereof – The finite meaning of one-one coordination of denumerable sets – The finite meaning of the diagonal procedure – 'Unordered' sets are arbitrarily well-ordered sets – The deceptive feature of geometric 'intuition' – The illusory procedure of forming the covering set of infinite sets (formation of power sets and the set of all sub-sets) – Inadmissible use of the concept of identity in forming the power set in *Principia Mathematica* – The sequence of types of transfinite cardinals – 'Calculating' with transfinite cardinals – Account of Cantor's theory of well-ordering – The definition of 'order' – Similarity and type of order – Definition of 'well-ordered' – Ordinal numbers – Segments of well-ordered sets – Cantor's thesis of the 'continuation of the sequence of natural numbers beyond the finite' – Cantor's fundamental series and limit numbers – The two 'principles of generation' for ordinals – Examples of ascent to higher ordinals – Hessenberg's order of natural numbers according to the order of type ω^{ω} – Epsilon-numbers – The two main theorems of the theory of well-ordered sets – The theorem of well-ordering – The series of aleph-numbers – Number classes – The continuum problem – 'Finitization' of the theory of well-ordered sets (ordinals) – Consecutive coupling and incapsulation of formation laws – The progression in the series of ordinals does not lead beyond the denumerable – The Löwenheim-Skolem theorem and its consequences – Criticism of non-predicative procedures – 'Self-transcending constructions' – 'Direct' introduction of higher powers – The meaning of the well-ordering theorem – The origins of the criticized doctrine in the psychology of cognition – Interpretation going beyond the mathematical procedure – Example: the theorem $c^n = c$ – Axiomatization of set theory – A representation of Fraenkel's axiom system – Remarks on various axioms (axioms of the power set, choice, unconditional existence) – The importance of the unexceptional discoveries of Cantor – The finitist tendency in modern foundational research.

VI. THE PROBLEM OF COMPLETE DECIDABILITY OF ARITHMETICAL QUESTIONS

Undecidability and being undecided – An example of the problem: Goldbach's hypothesis – Undecidability in polymorph (branchable) axiom systems – Monomorphism (unbranchability) of arithmetic – The theses concerning the infinity of proof procedures and the finitude of human thought – The illusion of four possible cases as to provability and definite decidability – Connection of this problem with that of the non-denumerable infinite (Brouwer) – Separation of the problem of definite decidabilitv from the 'decision problem' –

INTRODUCTION

When mathematical enquiry in the previous century succeeded in eliminating the concept of the infinitely small from infinitesimal analysis, this was regarded as a great advance in the sense of the postulate of purity in mathematical method.[1] For the prince of mathematicians, Gauss, no doubt under the influence of Kant's critique of reason, had rejected this concept as unmathematical, and most other creative mathematicians of the time could not ignore the fact that the infinitely small was a quite unwanted guest in the well defined domain of mathematical knowledge.

It was therefore with suspicion that, during the last decades of the century, people greeted the theses of Georg Cantor, who in his set theory undertook to establish a mathematical doctrine of the 'actual infinite', maintaining that the series of natural numbers had to be extended beyond the finite if one was not going to refuse acknowledgement to certain states of affairs that were quite within our logical grasp. L. Kronecker, above all, one of the most important mathematicians of Cantor's time, fought most resolutely against those theses.

The seemingly probative power of Cantor's arguments, however, carried the day against all objections, and since the turn of the century his 'set theory' has been regarded by competent mathematical enquirers as a legitimate mathematical doctrine.

What greatly contributed to this is the fact that Cantor's discoveries proved to be extremely fruitful for one of the most important mathematical disciplines, namely the theory of functions. On the other hand, disquiet was provoked by the fact that a rigorous working out of Cantor's principles leads to contradictions.

Yet the efforts of mathematicians – let us mention E. Zermelo and A. Fraenkel – succeeded or seemed to have succeeded in deriving set theory from a system of axioms in such a way that the contradictory ('paradoxical') sets dropped away, while those of Cantor's results that were regarded as legitimate were on the whole preserved. Nevertheless, the profoundest mathematicians could not fail to notice that the various axiomatic systems represent structures on unsafe ground; this state of affairs was recognized as unsatisfactory, and therefore efforts have never ceased to remove it.

As regards these problems there are currently three main opposed positions: the formalism of Hilbert, the intuitionism of Brouwer and Weyl, and the logicism of Russell. In my view the hardest stretch on the road to the solution of these problems has already been negotiated, although each of the three positions has captured only a part of the truth. That definitive clarity has not been reached I attribute in the first instance to the fact that a mathematician who has acclimatized himself in the symbolism of his science will find it peculiarly difficult to free himself from the spell of this symbolism and without prejudice to penetrate to the facts that this symbolism denotes.

In contrast we shall here show that one main origin of the problems concerning the 'actual infinite' (the transfinite in Cantor's sense) is to be found in the use of mathematical symbolism beyond its sphere of significance; and that if we eliminate its 'extravagant' use (in Kant's sense), namely a use no longer justified by the facts, then at once we can, without too much trouble, separate the wheat from the chaff, genuine knowledge from illusion.

Let us present a brief preliminary sketch of the general plan of this wrong path so dangerous for mathematical heuristics. We must distinguish three steps.

The first consists in some genuine item of mathematical knowledge. Linked to it there is secondly an interpretation that in a certain direction goes beyond the factual content and is expressed in a certain symbolism; thirdly, following that interpretation, this symbolism is in the end applied even where the objects of cognition initially denoted by it are lacking, which produces the illusion that there are such objects here too.[2]

If now criticisms are raised against this extravagant use of symbolism, the mathematician wrongly fears that the attack on his symbolism endangers those findings too which he rightly regards as established. Thus arises the strange notion that 'excessive rigour' is leading to an impoverishment of mathematical knowledge, so that we must guard against this demand becoming extreme.

In fact, however, to call a result obtained by lax methods 'knowledge' is justified only if we can reach the same result by unexceptionable methods.

In order to see this clearly, we shall in the sequel have to distinguish most rigorously between mathematical operations on the one hand and their interpretation on the other. It is in the operations alone that the cognitive content of mathematics resides, however heuristically important interpretation may sometimes be. As hinted above, it will then turn out that mathematicians, like enquirers in most other sciences, only seldom possess a thorough insight into the cognitive content of their symbolism; and that at precisely this point we find one of the most important psychological starting points for the problems of the transfinite.

These brief general observations will become fully clear only in terms of the investigations to be carried through below; but we may remark at this stage that the false argumentations outlined above further contain the starting points for metaphysical reinterpretations of certain facts that are fundamental for cognition.[3]

The preceding remarks suffice to suggest the close connection between the problems of the transfinite in mathematics and fundamental problems of epistemology. In the sequel we shall see in full clarity that important methodological problems crop up at the border lines of special sciences (here mathematics, although the same can be observed for other fields too), where they will be brought towards solution, or, if illusory, to dissolution, and that these problems are general problems of cognition. That is precisely why, in order to gain proper access to the basic methodological problems of the special sciences (and such, in the main, are the problems of the transfinite, too), we must start from considerations of the most general kind, from the basic facts of cognition. In the next section we shall exhibit these basic facts that must be clearly grasped for the purpose of our special analyses.

NOTES

[1] We must, however, not overlook the fact that Newton and Leibniz, the creators of infinitesimal calculus, were by no means unclear as to its finite character; witness the following passages:

"These ultimate ratios with which the quantities vanish are indeed not ratios of ultimate quantities, but limits to which the ratios of quantities vanishing without limit always approach, to which they may come up more closely than by any given difference but beyond which they can never go." Newton, *Principia Philosophiae naturalis*, Amsterdam 1723, p. 23.

"For it is in agreement with the geometry of the ancients to conduct an analysis of finite quantities and to investigate the first or ultimate ratios of nascent or vanishing finite quantities: I wanted to show that in the method of fluxions, too, there is no need to introduce infinitely small quantities into geometry." Newton, *Tractatus de curvata curvarum*.

"We must see whether we can demonstrate precisely . . . that the difference is not so much infinitely small but completely null, which will be shown if it is certain that the polygon can be bent indefinitely, to the point where even if the difference is assumed to be infinitely small, the error will be less." Leibniz, *Mathematische Schriften*, ed. C. I. Gerhardt, vol. 2, p. 217. The three passages are quoted in Latin in A. Voss, *Über das Wesen der Mathematik*, Berlin 1913, 2nd edition, p. 20 note.

However, Leibniz' notation, which was most suitable for heuristic purposes and for techniques of calculation, provoked the illusion of operations using infinitely small quantities. In the 19th century it therefore required the whole intellectual effort of Cauchy and Weierstrass to achieve recognition of the finite character of infinitesimal calculus. Operating with infinitely small quantities has moreover been influenced by metaphysical speculation, partly going back to Leibniz. Amongst modern philosophical tendencies the 'Marburg school' has drawn epistemological consequences from an infinitesimal theory working with the concept of the infinitely small, namely the theory of Hermann Cohen. That this theory is untenable is now hardly in doubt.

[2] Accordingly, we shall recognize in what follows that in some statements about the 'infinitely large' this concept, just like that of the 'infinitely small' in analysis, can be eliminated and replaced by finite concepts, without any change in the factual content of the propositions concerned. In other propositions, however, where the 'infinitely large' arises in talk of the 'non-denumerably infinite', such a 'translation' will not be possible. Such propositions will turn out to be sham judgments and will have to be eradicated from mathematics.

[3] Cf. Chapter I, note 12.

I

BASIC FACTS OF COGNITION

Adopting the 'common sense attitude' of experience and turning to events in space and time, we find a wealth of things with various properties and further we discover relations between the presence of things of the same or different kinds, in such a way that either both can be exhibited at the same time or in a certain temporal sequence.

Here the 'world' of which we become aware is taken as something existing independently of our own or anybody else's thinking, as something that is simply there and as such requires neither to be cognized nor suffers a change in its factual status or in its character by being cognized. Thus we think of the world as independent of the fact of certain thought acts aimed at it.

As soon, however, as we consider the act of thinking about the world by means of reflection, which is evidently possible, that which exists becomes a datum so that the question arises whether or how far within the given we can separate the aspects attributable to the subject (sense perception and reason) from those attributable to things.

That we must take subjective aspects into account at all was no doubt discovered only when, in the historical course of philosophic thought, the fact was noticed that the contents of perception are influenced not only by transformations in the object but also by changes in the mode of observation (position, coloured spectacles and so on). It is this that has given rise to over-hasty theses of the 'mere subjectivity of the world of the senses'.

Yet just as analysis leads from the object to subjective aspects, by showing that every determination of the object points to cognition and therefore being cannot be divorced from knowing; so likewise an analysis starting from subjective aspects leads back to the object, by showing that what we think about, the object of our thinking, is in a specific manner 'contained' in thought. This is the basic fact of intentionality, first clearly grasped by Franz Brentano, which must occupy a central place in any correct descriptive analysis of thinking, whether this be called phenomenology, descriptive psychology or epistemology.

The fundamental connection just identified, between knowing and being, may be put more precisely thus: every cognitive act aims at a state of affairs which is thought of as independent of that act, in the physical or mental (or psycho-physical) world. No state of affairs is thinkable that would be in principle unknowable.[1]

The first of these two observations is called the 'principle of the transcendence of cognition', and the second we shall call with O. Becker[2] the 'principle of phenomenological accessibility.' Let us make this clear by means of an example. Suppose, on the basis perhaps of an immediate visual experience, we judge that 'A few paces from here (where we stand) there is a house', this includes a thesis concerning the existence of the house, namely that 'it is there', whether it is being looked at or not; on the other hand it would make no sense to assert that the house exists but is in principle (not merely for technical reasons) unobservable. For judgments about its sensory qualities (figure, colour and so on) presuppose the possibility of an observer just as much as statements about the 'here and now' or 'then and there' of its existence, for this can be determined only relatively to the 'fixed' position of an observer.

With these two aspects, which we have marked above as points at which subjectivity breaks into the thinking process, we have by now indicated a basic division that runs through all thought acts. Following a customary terminology, let us denote this as the distinction between being-so and being-there.

Statements about being-so answer questions about the 'how', statements about being-there questions about the 'where and when' of objects (states of affairs). Determinations of the first kind may be linked with ones of the second in such a way that certain properties are asserted of spatio-temporally localized objects (states of affairs). This happens for example in the proposition 'Most houses currently standing in Vienna have brown roofs. To ascertain the truth of falsity of this assertion, we must make observations on certain individual objects (the houses of Vienna), and in order to test such an assertion it is essential to conduct the observations precisely on these individual objects. Not so with assertions concerning only being-so. Of course here too it is not in the least as though being-so were independent of being-there, but the latter is not specified; what remains undetermined is the spatio-temporal position of objects characterized as to their being so.

In order to grasp these matters, let us turn to the cognitive process performed when we make a statement about a certain sensory quality, for example a colour that we perceive at this moment on a coloured object in front of our eyes. Suppose we observe that colour can vary in three dimensions, namely in tone (red, blue, green and so on), brightness and saturation, then it is evident that this observation relates not only to the colour seen in the object directly in front of us, but that the same holds for any colour whatsoever. Thus we grasp an essential feature of colour in observing a particular object of a particular colour here and now; but that this object should be perceivable just here and now having this particular colour is a fact that does not enter the content of our cognition of the nature of colour. We abstract from it, but this abstraction does not mean that the nature of colour has a being beyond space and time and that 'alongside' or 'above' the colours occurring in real objects there exists a colour as such: it is merely that we note the invariance of particular aspects with regard to changes of a certain kind. Statements about the nature of colour holds for coloured objects wherever and whenever they may exist.

This fact is taken into account by the mode of expression according to which a statement about colour or quite generally about a quality is equated to the corresponding statement about all coloured objects or about all things of that quality respectively. We must, however, clearly understand that with such a qualitative statement the totality of the objects possessing the quality is neither given nor presupposed as given in it. Thus the statement by no means defines a totality of distinct things having this quality. We can, indeed, distinguish for any given individual thing whether or not it has that quality, but for a thing to be 'given' we need a principle of individuation.

A further example to elucidate this point so vital for our special investigations: if we speak of the inhabitants of London at a particular time, this automatically supplies a criterion by which we can in principle determine for any individual person, whether he falls under this concept, as in fact happens for example in a census; about London's inhabitants at a particular time we therefore can in principle obtain findings not already contained in the concept 'inhabitant of London'. For example, we can ascertain whether their heights all lie between 40 cm and 2 m.

If this is so, one says that all inhabitants of London are between 40 cm and 2 m tall, but this 'all' has not the same meaning as the same term where we spoke of all coloured things. To see this we must begin by observing that the former expression can have two different senses that are usually not kept apart. One is: 'no inhabitant of London is not between 40 cm and 2 m tall', or, what is the same: 'if something is an inhabitant of London, then it is between 40 cm and 2 m tall'. This proposition is true even if there is no inhabitant of London. In the second sense, however, the last mentioned negative or hypothetical assertion has added to it an existential assertion: 'there is an inhabitant of London', so that this compound assertion is false if there is no inhabitant of London.

Whether we use the first or the second sense, however, in order to decide as to the truth or falsity of such empirical statements we shall have to conduct enquiries about the existence of relevant objects, in order to ascertain either that they do not exist, or, if they do, whether all extant examplars satisfy the assertion in question. To return to the example, if there are inhabitants of London, then statements of this kind about 'all inhabitants of London' are in fact only collections of statements about inhabitants A, B, C and so on, about any and only such objects as fall under the concept 'inhabitant of London'. That such an empirical universal statement can be verified at all presupposes a spatio-temporal boundary or principle of individuation, as noted already and further to be justified below.

Correspondingly (and this determines the difference in principle between empirical and non-empirical universal statements), further statements about 'all coloured objects as such' in the sense of ones that simply follow from the nature of coloured objects (of colour) are impossible. For a question as to empirical existence is out of place here and in the absence of a principle of individuation we cannot analyse the universal statement into individual ones.

In the sequel we shall recognize that inadequate separation of empirical from non-empirical universal statements or existential statements is one main source of the difficulties arising in the foundation of set theory.

Within the scope of the epistemological discussions forming this section, however, we must begin by noting that the distinction just made is closely connected with the division by which all knowledge is divided into a priori and a posteriori.[3]

Modern philosophy has brought out clearly the character of this distinction, showing that the basic divergence between the two kinds of knowledge consists in this: a posteriori knowledge relates to the individual case as such, so that its validity appears in principle tied to that case; while for a priori knowledge the individual case by means of which we grasp the general feature, appears merely as an example for that feature and could be replaced by another example.[4]

In a priori knowledge we therefore abstract from all empirical facts, according to which things with particular properties (or with relations of a particular kind between them) exist at a fixed place and time; in this way properties and relations are isolated in thought according to their specific pecularities. We must, however, emphasize that to this mental process of isolation there corresponds no independence in the sense of Plato.

Thus, a 'red' cannot crop up independently, but only in a spatio-temporally extended thing. On the other hand, things are described by means of their properties, which are the stuff of cognition, predications about the subject-things. If then the individual thing is alone 'real', it is nevertheless built up of properties which therefore are simpler than it and amount to something logically prior with regard to it.

It is from this fact that all ontological enquiry has had to proceed; it is the main source both of the realist controversy whether priority belongs to real things or to kinds, and of the problem of substance. These were the problems with which ancient and mediaeval philosophy struggled and it was one of the aims of Kant's critique of reason to overcome them.

Dependence, however, belongs not only to properties in relation to corporeal things, but also figures on other levels of objectivity, with certain modifications.

Let us make this plain in terms of the above example. To every colour belong tone (red, blue, green), brightness and saturation, and these three aspects can vary independently of each other: every tone can appear at any level of brightness and saturation, nor do these last two aspects determine each other. A colour with its particular tone, brightness and saturation is given in perception as a single phenomenon; it is only when we analyse this, by reflecting about the variations in the different directions, that we are led to the three aspects mentioned, which are considered separately although they cannot exist separately.[5] On the other hand, they are the simpler items, since the phenomenon of colour exhibits them in combination so that they stand to it in the relation of elements to compound. Such a relation must, however, not be viewed as a summation. Quite apart from the fact that summation correctly applies only to numbers, we here have not a mere setting alongside but a genuine 'inside one another', a synthesis, for the individual aspects form a phenomenon of a higher complexity in a completely different way than adjacent pieces of matter form a corporeal thing. As emphasized before, these relations are fundamental for the construction of the world of objects and they reach far into the problems of the individual sciences. It is odd that the precise and evident formulations on this topic due to E. Husserl[6] have hardly yet been applied in methodology, so that there is still a bitter struggle about basic positions concerning which theory has already spoken the decisive word.

Let us now cite those of Husserl's observations that are most important for us. The central concept here is that of foundation, which Husserl defines thus[7]: "If according to the law of its nature an α can exist as such only in a comprehensive unity linking it with a μ, we say that an α as such needs founding by a μ, or that an α as such requires completing by a μ". "If then α_0, μ_0 are realized in a whole as particular individual cases of the pure kinds α and μ standing in the relation stated, we call α_0 founded by μ_0; exclusively by μ_0 if the need of completion is fulfilled by μ_0 alone. This terminology can of course be transferred to the kinds themselves, an ambiguity which is quite harmless here. Less definitely we further say that the two contents (or pure kinds) stand in a founding relation or in the relation of necessary connection; it remains of course an open question which of the two possible and mutually non-exclusive relations is meant. The indefinite expressions: α_0 requires completion, it is founded in a certain aspect, evidently mean the same as the expression: α_0 is dependent".

"Every part that is independent with regard to a whole we call a piece, and every part dependent on the whole an aspect (an abstract part) of this same whole; it being indifferent whether the whole itself, viewed absolutely or relatively to a higher whole, is independent or not. Abstract parts can thus in turn have pieces, and pieces abstract parts. We speak of pieces of a duration, although this is something abstract, and likewise of pieces of an extension. The forms of these pieces are abstract parts contained in it."[8]

"An object in relation to its abstract aspects is called a relative concrete".[9]

"A concrete that is not itself abstract in any direction, can be called an absolute concrete."[10]

"Talk of the uniformity of foundation signifies that every content is connected with every other, by foundation, either directly or indirectly. This can occur in such a way that all these contents are immediately or mediately founded in each other without external help; or in such a way, that conversely all of them together found a new content, again without external help."[11]

Let us now build further on these observations of Husserl. To begin with we directly infer that a corporeal thing is not 'alongside' its properties but 'in' them, although they are not isolated, but only capable of being considered in isolation. The possibility of thus isolating them in thought is, however, based on free variability within certain limits, as we have shown. For example, colour and extension belong inseparably together, but nevertheless can be considered each on its own, because any arbitary colour is compatible with any arbitrary extension. If, however, we hypostasize this isolability into independent existence, then we become involved in boundless metaphysical speculation.[12]

Let us pursue this example, in order to ascertain further important facts. We have earlier described a priori cognition of being-so in outline, and recognized that what characterized it was the fact of grasping the nature of a kind by means of an arbitrary exemplar. This grasping can, however, occur on different levels of generality. For example, a thing of a certain shade of red can allow us to treat as given either this shade, or the being of 'redness' or finally the being of 'colour'. Here lies the boundary of this kind of generalising. Following Husserl, we may therefore call colour a highest kind.

Husserl defines this concept, which is of basic importance for logic, in the following passage, where moreover he fixes the concept of eidetic singularity, likewise of great importance for our investigations:[13] "Every being, whether with content or empty (that is, purely logical), takes its place in a sequence of levels of being, a series of levels of generality and specificity.[14] To these series there necessarily belong two boundaries that never coincide. Descending we reach the lowest specific differences, or as we say, eidetic singularities; ascending through general kinds we reach a highest kind. Eidetic singularities are beings that necessarily have above them beings 'more general' than their kinds, but no further specialisations below them, in relation to which they would themselves be kinds (adjacent or mediately higher kinds). Likewise, the highest kind is that above which there is no further kind."

Accordingly, the highest kind 'colour'[15] has as its eidetic singularities (last specifications in the sphere of being-so) the particular shades ascertainable in individual objects (a red like that of some particular roof).

Husserl's distinction (likewise highly important for our subsequent analyses) between 'with content' and 'empty' or 'formal' rests on the presence or absence of specific material of sensation (sense data). According to this, order and number, for example, are formal objects.

That a kind B is lower than a kind A (that it 'falls under kind A') means simply that an arbitrary individual thing that falls under B also falls under A. The relation between a more general and a less general kind or an eidetic singularity should therefore not be misconstrued by equating it with the relation between a kind and the individual things that fall under it.

Let us make quite clear what is the difference of principle between these two cases. That a particular spatio-temporal something, a 'this-here', has particular properties, is an empirical finding: this-here is red, but it is theoretically possible that the same location in space and time be occupied by something that is not red. However, that a thing with a particular shade of red is a coloured thing is not something that could now be and now not be, because with the concept 'particular shade' we already think that of colour as well. Thus every coloured thing has a particular colour and a particular shade of it. That in spite of this we can make statements about 'colour as such' (about 'everything coloured') rests on the empirical fact, described at the beginning of this section, that abstraction comes on various levels; this enables us to use a given being-so (a given concrete thing) to ascertain qualitative features that agree only in part with those that are given. Statements about a 'general' being-so (redness, colour[16]) are thus about being-so that is only partly defined. Specification is therefore closer definition of being-so, and finds its conclusion in the complete particularity of eidetic singularities. In order to understand the problems of the transfinite, too, it is important to see clearly that there are no different layers of generality of what is empirically given as being-so, but that every being-so is given in an eidetic singularity. For we shall recognize that every particular natural number is a last logical specification, that is a formal eidetic singularity in Husserl's sense. In particular, this insight will prevent the fatal merging of properties or sets of concrete individual things with 'properties' or 'sets' of numbers.

Concept formations of the latter kind are indispensable in the attempt to ascend to the non-denumerably infinite, whose logical legitimacy lies at the centre of methodological controversy in mathematics; let us therefore analyse it a little more closely. To this end we must first consider the empirical assertion that a particular thing T has a particular property P. What interests us is the question, in how far the thing T_1 can count as definite independently of having the property P_1. (That it is definite we presuppose, since our assertion was assumed to be empirical, so that P_1 cannot be analytically contained in T_1). We now recognize that what here produces unambiguous particularity is the principle of individuation. Our proposition thus means: this object here and now (or there and then) has the property P_1. It can be characterized besides by other properties that it has, or by relations it bears to other beings, but such need not be the case, for the principle of individuation is a necessary and sufficient condition for its unambiguous particularity.

What then about the assertion that a particular property P_1 (for example a particular colour) has a particular 'property' P_2 (say, a particular brightness)? Here it turns out that a property (colour) P_1 can count as particular only if it has been ascertained whether it does or does not have the property P_2 (a particular brightness). Accordingly, statements about 'properties' of properties are mere linguistic transformations of statements about the properties themselves and must therefore be eliminable if we introduce a suitable symbolism. Otherwise we are confronted with sham statements, namely a sequence of words (symbols) that at first blush seem to have sense but turn out, on closer analysis, to lack it.

Another possibility of introducing 'properties' of properties seems to lie in the fact that when things with a particular property exist (can be exhibited at particular places and times), this is described as a 'property' of that particular property. For example, that a particular room at a particular time contains three yellow things would then be described as a property of the colour 'yellow'. Yet this view, prevalent in mathematical foundation research, cannot stand up to critical reflection. In order to appreciate this, we need merely become fully aware that every 'property' is the result of a process of abstraction in which we set aside the principle of individuation in particular. Spatio-temporal indefiniteness therefore belongs to the meaning of any property as such, so that it would be senseless to construe spatio-temporal determinations as 'properties' of properties and to anchor this construal in a logical symbolism. Still, what is meant by saying 'it is a "property" of the colour yellow that three yellow things are in a particular room' has a proper sense; namely no more and no less than 'in that particular room at that particular time there are three yellow things'. (The question how we are to define thinghood is here of no account.)

In general we therefore have the following: wherever people talk of 'properties' of properties, 'properties' of 'properties' of properties and so on, the word 'property' has no uniform meaning: such propositions therefore have a sense only insofar as they are 'translatable' into other propositions in which the word 'property' is not used in this twofold manner.

What has been said above about 'properties' of properties and so on can likewise be transferred to 'properties' of numbers and so on. Here too, for reasons analogous to those above, the statements are always about numbers themselves; the contrary illusion is to be sought in defects of symbolism. As a result of this insight we must eliminate the so-called extended functional calculus, which is closely linked with the most important foundational problems of mathematics. This will become clearer still in the next two sections.

We have by now also obtained the basis for ascertaining the sense of the 'there is' propositions, infelicitously called 'existential judgments' (propositions of the form 'there is an S for which P holds') in the sphere of the pure statements of being-so, that is a priori propositions. Here 'there is' amounts merely to the empty form of a general being-so that is to be further specified.[17] For what follows, it is very important to be clear that neither does the concept 'all' here encompass an infinite set nor the concept 'there is' denote a selection from an infinite set.

The erroneous view that assertions of mathematical existence are independent judgments not requiring completion is closely connected with considerations touching the sphere of the non-denumerably infinite. It is one of the main tasks the present work sets itself, to refute the assumption that there is such a sphere and thus to remove the strongest motive for severing 'mathematical existence' from 'constructivity'.

Connected with the concept of a highest kind there is also the following incompatibility relation which is vital for all cognition: different species of the same highest kind cannot be united. Thus, while a spatio-temporally bounded thing that is unambiguously determined may well be blue and rectangular or blue and round, it cannot at the same spot be both blue and yellow.

Having so far concentrated mainly on the cognition of being-so, we must now turn to the principles for cognizing the being-there of individual things. Here the highest principle is that every object (state of affairs) unambiguously fixed as to spatio-temporal position is also unambiguously determined as to its being-so.

The content of empirical knowledge, however, is the coordination between being-there and being-so, that is the answer to the question 'how' concerning the objects (states of affairs) at a particular place and time.[18] It turns out that about this we can often make rather reliable predictions; that is, assumptions about particular constellations of events hold with considerable accuracy.[19] The objective presupposition for this are repetitions in the course of the universe, that is the return of similar constellations or aspects thereof.

This 'fact', whose continuance beyond the present can of course only be 'assumed' (surmised), is the basis of the cognitive content of universal empirical judgments, such as 'all viviparous animals breathe through lungs'. This judgment is an assumption whose complete verification is in principle possible only in such a way that, if there are viviparous animals at all, each of them must be examined as to whether it does breathe through lungs. This in turn is in principle possible if the spatio-temporal domain containing viviparous animals is in some way bounded, for otherwise the task of ascertaining their breathing habits is not definitely formulated.

In universal empirical judgments, so far as they are fully made rational so that their sense is perfectly clear, a particular spatio-temporal domain is always understood as well; say, the 'Earth', in our example. This holds even more where the object of statements consists in relations of succession between events, namely in causal judgments. Indeed, the assertion that events of kind A_2 'generally' follow on events of kind A_1 is decidable, even as to validity of the individual instance, only by bounding the spatio-temporal position of A_2 relatively to A_1; that is, we must indicate a spatio-temporal neighbourhood of A_1 within which A_2 occurs. Otherwise a refutation of a causal assertion for a particular instance would be in principle unthinkable. It is enough to recall the content of any precisely formulated law of physics to grasp this point.

The case is quite different for pure statements of being-so, as we noticed at the beginning of this section. They hold not for a particular spatio-temporal domain but for any arbitrary such domain, that is for 'all' domains, since they are verified or refuted not by examining a totality of real objects with the property concerned, but can be checked on an arbitrary object that possesses the property.

Let us distinguish the 'universality' of empirical judgments from that of pure judgments of being-so, and with Husserl,[20] call them 'individual universality' and 'specific universality'.

That this radical divergence, which we have already analysed in detail, was usually overlooked, has been fatal both for logic and for foundational investigations in mathematics. For logic, we shall make good this claim in the present section, and for mathematics in subsequent ones. In analysing foundational questions in mathematics we shall see that in the most important places (namely in the problems concerning natural numbers or complete induction, as well as in those concerning irrational numbers and the non-denumerably infinite) the main nodal point of confusion is formed by the interlacing of these two heterogeneous aspects of the concept of class (or set), so that in dissolving this equivocation a good many of the consequent difficulties are overcome. The merging just mentioned has two main roots: the first lies in not sufficiently noticing the fact that to every empirical judgment there belongs a definite spatio-temporal domain for which it is to hold; for by ignoring such a domain we create a dangerous linguistic blurring of the distinction between the 'universality' of empirical judgments on the one hand and that of non-empirical ones on the other. The second and perhaps even more important main root of this confusion lies in the sensualist theory of abstraction (see note 5). This completely fails to capture the sense of general universality and is therefore in particular incapable of correctly grasping those last formalising generalisations that lead to the sphere of logic. To this latter we must now return.

Logic is sometimes defined as the doctrine of the formal criteria of truth. In order to gain some insight into its nature, let us try to clarify the sense of 'truth' and of the 'formal', which will indeed show that this definition is not unobjectionable. Let us begin with 'truth'.

The customary view concerning the relation between the three concepts 'correctness of individual acts of judgment', 'truth' and 'being' is this: 'A judgment is "correct" if the assertion it contains is true; whereas an assertion counts as "true" if, in ways to be further specified, it agrees with what is.'

However, this formulation is not unobjectionable, since between the fact of correctness (accordance with what is) of judgment and the being to which they correspond, it interposes a further intermediate concept, whose relations to the other two is not clearly grasped. To define the concept of truth of judgments correctly, we should say rather: 'a particular judgment is true' means 'whoever at whatever time and place makes this judgment, judges correctly (in accordance with the facts); he asserts what is the case'.[21] (The criteria by which in any given case correctness is determined can here be left aside.) Thus the concept of truth of judgments gives expression to the fact that their correctness is invariant with respect to changes in their factual setting, namely in the person who judges and the place and time of his judging. The origin of this invariance lies in the independence of being from the fact of its being thought on various occasions, as we observed at the beginning.

The state of affairs here described was then reinterpreted in a way suggesting that, alongside what is, there existed a 'realm of truth', a judgment being correct when it falls into that realm. However, as just observed, this interposition is unjustified: there is no 'third realm' of holdings that link being and thinking. 'Judgments as such' and 'truth as such' do not stand beyond or 'above' psycho-physical subjects and objects about which judgments are made, but they are merely terms designed to emphasize a judgment's invariance with respect to changes of judging persons and its holding for arbitrary objects of judgment.

What has just been said makes clear what is meant by a 'judgment in the logical sense'. An act of judgment is to think of a state of affairs as existing. If what we take into account is merely the state of affairs being thought (the object of the judgment) and the characteristics by which it is thought (the content of the judgment), while abstracting from 'occasional data' such as who judges and when and where, we obtain a 'judgment in the logical sense'.[22]

This insight now enables us to render innocuous an ambiguity in the term 'judgment', namely between act of judgment and object of judgment. Evidently the fact of thinking of a state of affairs and that state of affairs itself must be sharply distinguished. However, in making this distinction we have not yet removed the danger of merging disparate aspects in this area: that requires the further distinction between act and object of judgment.

To make this difference more precise, let us refer to E. Husserl's fundamental investigations on this theme and quote his own words: "Every expression not only says something but also speaks about something; it not only has a meaning but also refers to some objects or other. This relation may well be multiple for one and the same expression, but object and meaning never coincide. Both of course belong to the expression only through the mental act that gives it a sense; and if in respect of these 'conceptions' we distinguish between 'content' and 'object', this means the same as distinguishing in an expression that which it means or 'says' and that about which it says something."[23]

"Names give us the clearest examples for separating meaning and reference to an object. As regards the latter we here are wont to speak of 'naming'. Two names can have different meanings but name the same thing. For example, the conqueror of Jena – the conquered of Waterloo; equilateral triangle – equiangular triangle. The meanings expressed in each pair are obviously different, although the same object is meant in each. Likewise for names that, because of their indeterminacy, have an 'extension'. The expressions 'equilateral triangle' and 'equiangular triangle' refer to the same object and have the same extension of possible application."[24]

In order not to clash too much with the terminology customary in foundation research in mathematics, we shall in what follows denote by the 'reference' of a proposition the state of affairs asserted in it, while what Husserl here calls 'meaning' we shall call 'sense' or 'content'.

A further important fact about 'judgment' concerns the relation between its sense and its truth.

A judgment is true or false, according to whether the state of affairs asserted in it does or does not obtain, and it has a clear and unambiguous sense only if it is certain according to what criteria this 'obtaining' or 'not obtaining' is determined. For otherwise it would remain unclear what is to be understood by 'obtaining' and 'not obtaining': a judgment that has sense must therefore be either true or false and it must be certain under what conditions (on the basis of what criteria) it is to be regarded as true and under what as false.[25]

This observation can be extended to tautologies and contradictions as well (see p. 36ff); here the 'criteria' lie in the fact that, after suitable transformations, tautologies and contradictions must 'show' themselves.[26]

The above analyses immediately give rise to important implications for a series of problems that play a big role especially in symbolic logic and hence also in the logicist theory of mathematics. The first thing we must consider here is the logical treatment of statements about statements, statements about statements about statements and so on.

Take the two propositions 'all men are mortal' (p) and 'all negroes are mortal' (q) and observe that between them there is the relation 'p implies q', which is to say that from the truth of p that of q follows. This seems to be a statement about statements.

However, on the basis of what we have just learnt about the concepts of 'truth' and 'judgment', we see that this formulation merely represents a transposition of judgments about states of affairs into 'judgments about correct judgments about states of affairs'; for that from the truth of p that of q should follow, it is a necessary and sufficient condition that the state of affairs denoted by p should include the state of affairs denoted by q, that is the state of affairs q always holds if the state of affairs p does hold.

This transposition thus alters the content of the judgments, since they now relate to acts of thought in general (namely by arbitrary persons at arbitrary times and places), but every judgment is true or false together with its transposition – they are 'equivalent', as we say for short. The logical 'relations' between two judgments are thus completely determined by the relations between the states of affairs on which the truth or falsity of the judgments depends, and this holds in the same way for judgments that relate directly to those states of affairs as for judgments about such judgments, and so on indefinitely.

The logical 'relation' ('truth relation') between judgments is therefore not something that exists alongside the criteria residing in the corresponding states of affairs. What produces a semblance to the contrary is mainly that object of judgment and act of judgment are not sharply distinguished. For while from the assertion 'whoever, at whatever time and place, judges in a particular way, judges correctly', we can, by filling the empty personal and spatio-temporal places with empirical data, make empirical statements about the correctness of empircally determined acts of judgment, nevertheless we recognize that such statements no longer directly relate to 'judgments in the logical sense'. Every statement about judgments or about their truth must therefore be translatable into a statement about states of affairs that form the object of those judgments. That is why there is no need to introduce a symbolism for judgments about judgments (the extended functional calculus). Nor is it proper to justify such an extended calculus by advancing the argument that the truth of a judgment is denoted as a 'property' of it, for the analysis of the concept of 'truth' just carried out has made plain how this way of talking is to be understood and correctly resolved. From this it follows that it is incorrect to denote truth as a 'property' of judgments.

Our observations further yield consequences for the problems of so-called non-predicative judgments, namely 'judgments that contain themselves'. For if we here distinguish clearly between act and object of judgment, we recognize that the question whether a judgment can contain itself is correctly formulated thus: 'can an act of judgment be meant along with the corresponding object of judgment?' With this formulation, however, a considerable part of the difficulties arising in this problem area has already been cleared away. Since non-predicative judgments are one of the main sources of logical antinomies, we shall have to have a closer look at these in the last section, which deals with their analysis.

Finally, for those familiar with the theories of Frege and Russell, let us point out that, with the correct distinction between content and object of judgment, the controversy about the extensionality thesis[27] is resolved in its favour insofar as a propositional function ϕ figuring within a statement can be replaced by any formally equivalent propositional function ψ without the statement changing in truth value. In the statement $f(\phi x)$ we can therefore replace ϕx by its extension.

If, however, as is usually done in symbolic logic, we link the concept of extension of a propositional function with the idea of a 'totality of objects satisfying this propositional function', there arises the objection raised above against the merging of individual and specific universality, to which we shall return in the analysis of logical relations between concepts which follows.

Just as a judgment arises from thinking that a state of affairs holds, so a concept arises from intending[28] an object (state of affairs). Here, however, we find further complications as a result of the fusion of logic and language.[29]

This fusion stands out with particular clarity in the conception of definition according to which its logical content amounts to no more than the substitution of one name for another.[30] If this were correct, every translation from one language into another would have to be called a 'definition'. Evidently, however, when we regard definition as a logical procedure we are not including translation. In fact we mean by the definition of a concept the indication of how an object (state of affairs) is made up of other independent or dependent objects (states of affairs). Here 'make-up' is to be understood firstly as the constitution of more complex objects in terms of simpler ones; and secondly as the quite different specification of general objects (definition of species by means of genus). Accordingly, definitions that aim at empirical objects have a sense if and only if they represent compositions of compatible empirical data. Whether exemplars of the kind defined can in fact be pointed to (houses) or not (centaurs) is a question that does not concern logic.

This settles the question of the connection between definition and the existence of what is defined. In defining empirical concepts[31] we think spatio-temporal (or simply temporal) data (that is, those whose being-so is evident) are given; but this leaves undecided whether such a combination does in fact exist somewhere in space and time.

This last question, however, simply vanishes where we define formal concepts as in logic and mathematics. The indissoluble connection with the world, which excludes a logical 'initial creation' by means of 'creative definitions', nevertheless becomes evident here too, as soon as the meaning of formality is properly grasped.[32]

Of special interest to us is the theory of denotation and connotation of concepts, since here the fusion of individual and specific universality just exhibited plays an essential part that is equally fatal to logical, philosophical[33] and mathematical problems.

The misconception in question gains expression both in the definition of the logical denotation of a concept as the 'set of objects falling under it' and in the definition of the connotation of a concept as the 'sum of the object's characteristics'.[34]

Without as yet giving an unambiguous account of the nature of logic, we can assume as certain, that logical relations do not depend on changes in the real world (which encompasses physical and mental facts),[35] whereas on the definition of denotation just mentioned the opposite would have to be case. What we have here are on the contrary only relations between kinds of things, the real existence of individual cases being of no account.[36] Even apart from this, the formulation is misleading. It is not the case that there is firstly a concept, secondly its denotation and thirdly its connotation, but the connotation is simply the concept itself while its denotation is the sphere of objects to which it refers, which latter is completely determined by the connotation.[37]

The assumption that connotation is something that belongs to a concept without being identical with it can be traced back above all to the identification of a concept with its sign (word). Making denotation independent from connotation, however, arises mainly from misconstruing the 'quantitative relation' between the two, according to which every concept has a denotation and a connotation of determinate sizes varying inversely with each other. Yet, in fact, the denotations and connotations of two arbitrarily chosen concepts need not be comparable as to their sizes. Two concepts C_1 and C_2 can be thus compared only if C_1 refers to every object to which C_2 refers, or conversely, for any arbitrary constellation of the facts. If these relations are not reversible, we say in the first case that C_1 has greater denotation and smaller connotation than C_2, and conversely in the second case; if the relations are reversible, C_1 and C_2 are called equal in denotation.

The assertion that the denotation of a concept is greater than that of another, and the assertion that its connotation is smaller than that of that same other concept (or conversely), thus describe the same state of affairs: they mean the same and merely say it in different ways.

This raises the following question which is fundamental for an understanding of logic: are logical propositions all of such a kind as that concerning the relation between denotation and connotation of a concept? Does the cognitive content of logic exclusively consist in the fact that it reveals as irrelevant those aspects that are connected with the mental fact of thinking as something taking place in time, starting from certain premisses and leading to certain conclusions? Is logic perhaps merely a set of rules concerning the use of certain symbols?

Let us clarify the problem by starting from the example of a proposition '*p* implies *q*'. This can be replaced by the proposition 'not-*q* implies not-*p*'; that is, both propositions mean the same.

What, then, is the difference? From modern logicians who are close to mathematics we may expect the reply that the difference lies in language; but this answer needs to be made more precise in any case, since the two forms evidently do not differ in the way in which an English formulation differs from the corresponding French one. Rather, we have here a variation in the thinking itself, quite apart from linguistic formulation. This assertion, too, needs making more precise, which will above all have to bring out the sense of formal concepts like 'not', 'or', 'implies', since we are here dealing with transformations.

The following conception here suggests itself: in the world there are only objects (states of affairs), but not negative objects (states of affairs); therefore negation does not say anything about the world; similarly it might be said that the existence of a state of affairs *A* is a statement about the world, and likewise the existence of the state of affairs *B*, but nothing in the world corresponds to 'and' in the statement '*A* exists and *B* exists'. Likewise for disjunction and implication.

This view, however, is not correct, as we shall illustrate in the case of negation and conjunction. Let us begin with negation: to grasp its sense, we must first be clear that no (complete) affirmative empirical judgment simply asserts the existence of states of affairs, but that it always must trace out a spatio-temporal or personal and temporal domain, as we have observed before. Accordingly, the negation of such a judgment means that, within that finite domain, objects (states of affairs) of the kind in question do not occur. Such a statement, however, belongs just as much to the description of that domain as an affirmative statement; for in order to know the domain it is not enough to know that states of affairs *A*, *B*, *C* . . . do occur in it, but we must know also that other states of affairs *P*, *Q*, *R* . . . do not occur in it.[38] The ideal of complete knowledge of the domain consists in knowing of every state of affairs occurring there that it so occurs and in knowing further that no other state of affairs does. Negation is thus an essential element in the description of the world, and therefore an element of the world itself.

That conjunction also is linked with the world is evident if we grasp that every complete statement about empirical existence includes a spatio-temporal or personal and temporal determination. For we then recognize that the conjoined existential statements 'there is a state of affairs S_1 and there is a state of affairs S_2' can be transformed in such a way as to become 'a state of affairs S_1 occurs along with a state of affairs S_2 in the same bounded domain'. Thus 'and' here too means community, sameness in a certain respect, just as in the case of community of properties (where this is evident at once).

This cognitive fact has been somewhat obscured by the fact that we mostly do not operate with the states of affairs themselves, but mentally with the statements, understanding by '*p* and *q*' the 'being true together' of *p* and *q*, although this means no more than '*p* is true', '*q* is true'; for, as we have observed, it is a mistake to envisage truth as a property of judgments. Therefore we cannot assemble a plurality of judgments on the basis of their 'common property' of being true.[39]

That disjunction and implication are linked with the world just as negation and conjunction are, we have already emphasized. This results directly in any case from the fact that the former pair can be reduced to combinations of the latter, as is shown in symbolic logic.

That this holds, however, seems incompatible with the fact that we can operate with signs, named 'not', 'and', 'or', 'implies', on the basis of certain rules, without calling upon the sense of these signs as just exhibited, while nevertheless all logical propositions about those concepts find their one-one correspondence in this formalism.

Yet this is merely an illusion of incompatibility, as is shown by the following reflection: that from a statement *p* a statement *q* 'follows' means that the state of affairs whose existence is expressed in the statement *p* includes the state of affairs asserted in *q*, so that in *p* the state of affairs asserted in *q* is asserted as well. As to the rules fixed for the use of the signs employed, we must understand them as follows:

(1) That for one formula we can substitute another signifies that if we interpret their contents they both are to mean the same.

(2) That from one formula we can deduce another signifies that if we interpret their contents the latter is included in the former.

As regards the signs constituting the formulae we here presuppose that the same sign means the same thing, that is denotes the same thing irrespectively of the way the facts happen to be (the state of the world). Now these logical conventions of sameness and inclusion of meaning, which find partial expression in the initial formulae, are so chosen that they 'fit' the logical concepts of negation, conjunction and so on; that is, the conventions are a model of the mental operations performed with these relations, so that in the corresponding 'metalogical' or 'metamathematical' interpretation everything must be in order.

In order to grasp the sense of what is formal, or of logic, we must therefore distinguish the following. Firstly, there are statements that have sense for arbitrary states of affairs ('states of affairs as such'); for example, one can deny any arbitrary state of affairs and assert or deny for states of affairs the holding of shared features to be further specified (sameness in a certain respect). The point of view that takes into account those and only those aspects that do not relate to factual particularities constitutes the sphere of the formal, or, as we might say, the sphere of logical concepts.

Secondly, we have seen that the same formal connections can be thought in different ways, the variance arising from the different mental starting point (as in 'p implies q', 'non-q implies non-p'). That these two statements signify the same becomes clear if we reflect on what each of them says, provided of course that the linguistic signs have their usual meaning. The result of this reflection here lies in the insight that certain data belonging to the process of thinking are objectively irrelevant. Logical propositions thus assert nothing about the world but declare the total or partial sameness in meaning of assertions about the world.

Thirdly, logical transformations, which are legitimized by reflection on the meaning of formal concepts, can be semiotically fixed; we then need no longer refer to the meaning of the signs with which we operate. However, transformations as such are not thereby deprived of sense; for 'deducibility' of one formula from two others means that if we interpret them as to content the meaning of the first is contained in that of the other two.

In operating with formulae we thus do not attribute meaning to them, but the rules of operation do determine relations between their meanings when we interpret them as regards content.

Since we denote as 'formal' or 'logical' not only the problems mentioned under the first and second points above but also those mentioned under the third, there arises a danger of confusing ambiguities which we must take care to avoid in what follows.

We can now recognize without difficulty what is the significance of contradiction in logic. That one 'cannot', that is 'must not', assert and deny the same state of affairs, means that in doing so one asserts nothing at all. Where we have a genuine assertion, it never contains its own negation, which therefore cannot be inferred from it either, for 'inferring' from certain propositions (premisses) to other propositions signifies simply the exhibiting of meanings that are already contained in the premisses, as we have just explained.

In a system of logical formulae in which we abstract from the meaning of the signs, the above impossibility must reveal itself in that formulae which would negate each other when interpreted as to content cannot both be derivable.

The preceding analysis of logical inference, however, does not agree with the traditional view insofar as the latter takes inference to exhibit a connection between truths, while we declare it to be exhibiting a connection between meanings. That this latter view is the correct one is easily seen if we consider inferences from hypothetical premisses. For the assertion 'if *p* is true then, logically speaking, *q* is true', evidently requires justification in terms independent of the truth criteria for *p*, since it is not a question of empirical connections. It thus emerges that *q* must be true because, if *p* is true, *q* asserts[40] nothing that is not also asserted in *p*. The logical connection between statements is thus a connection between meanings, where a statement's meaning must be understood as that which it asserts. Accordingly, false statements too may imply true statements, since in false assertions correct material may be asserted as well.

What, then, of the widely misconstrued proposition of symbolic logic, that any arbitrary false proposition implies any true and any false proposition? For example, the proposition $2 \times 2 = 5$ would imply both the judgments 'blood is red' and 'blood is green'. This logical proposition must be understood as follows. If we construe logical connection as a truth connection, then '*p* implies *q*' means no more than the negation of '*p* and not-*q*'. Since, however '*p* and not-*q*' is always false when *p* is false, 'not-(*p* and not-*q*)' is then always true. On the other hand, 'not (*p* and *q*)' is also true, which in logistic terminology corresponds to the statement '*p* implies not-*q*'. On this view we can therefore derive any proposition at all from the contradictory proposition '*p* and not-*p*' which is 'always false'. Without using the thesis that logic is a matter of truth connections, one might try to justify the view just stated by saying that the proposition '*p* and not-*p*' asserts everything so that everything is derivable from it. However, this would not hold water: for a contradiction asserts nothing at all, it offends against sense.

In general we may say that the concept of implication stipulated in the terminology of symbolic logic merges in a misleading manner specifically logical implication with empirical 'implication'. In the latter, an assertion of implication is indeed established or refuted by ascertaining certain states of affairs (or, to use the appropriate technical language, by ascertaining the truth of certain judgments); but this does not hold for logical implication, which is a pure connection between meanings.

The insights gained now enable us unambiguously to fix the sense of logical propositions and concepts. As we have seen, it is characteristic of logical propositions that they make no assertions about the world but merely make clear how one and the same matter can be differently thought or expressed: logical propositions are tautologies.[41]

Is it also true that every tautology is a logical proposition? For example, are we to say that a proposition like 'snow is white or not white' is a proposition of logic? Not if we are abiding by what for over two thousand years has been understood, if often imprecisely, by logic; for that proposition contains some concepts that are not logical. Therefore only such tautologies as contain no concepts other than formal ones are to be called logical propositions.

We have already observed that concepts are to be denoted as 'formal' if they contain no reference to facts of sensation; now the further question arises, whether we can precisely describe or delimit the content of meaning that exists in the formal domain. That this can be done is suggested by the fact that in existing symbolic systems, above all in that of Principia Mathematica, logic can be constructed with the help of a few basic signs.

To see this more clearly, we must remember that there are only three basic cognitive forms, which acquire their content according to the relevant factual properties of the objects being considered.

(1) Arbitrary states of affairs can be thought of as not holding (can be negated).

(2) A plurality of arbitrary states of affairs can be examined as regards the existence or non-existence of certain common features.

(3) Every property of objects occupies a certain place within a graduated sequence of generality.

These three cognitive schemata form the basis of the formal domain and every symbolic representation of logic must respect them as guidelines. The three schemata respectively correspond to the words 'not', 'and' and 'all' (in the sense of specific universality). As Sheffer[42] has shown, negation and conjunction can be replaced by the relation 'incompatible with', if negation is denoted as self-incompatibility.

In this catalogue of basic relations one indispensable concept seems to be missing, namely that of identity. Deeper analysis, however, shows that the exclusion is justified; for 'identity', whether as such or filled in by contents whose formal schema it would represent, is not a concept that belongs to the world (that is, essential to describing the world). Indeed, several objects can never be identical, otherwise they would not be several, while it makes no sense to say that an object is 'identical with itself'.[43] The concept of identity is used sensibly only where we wish to express that the same matter is being thought of in different ways, or being denoted by different signs.

For what follows the most important application of this insight is the observation that the sign of equality in mathematics relates not to the mathematical objects (numbers) themselves, but to mathematical symbols. That we put an equality sign between two mathematical symbols says that they mean the same thing.[44]

If by way of conclusion and summary, we are to specify the relation between logic and the world, we can say that while logic says nothing about the world, it presupposes the world.[45] For as we have seen, logical concepts constitute the schemata for describing the world. Moreover, the analysis of connections between the meanings of statements, which is a main theme of logic, presupposes an existence which as object of these statements constitutes their meaning. From this point of view we can understand the historical position of logic as theory of science, a matter which we cannot pursue in detail here.

Clarity as to the sense of logic is essential for any deeper grasp of the problems in the foundations of mathematics, as we shall see especially in the subsequent investigation on the concept of number.

Before tackling that problem it will, however, be apposite to undertake a brief analysis of the language of mathematics (symbolism) on the one hand, and of the axiomatic method on the other, since an understanding of the roles of both within mathematical knowledge is important for grasping the character of mathematical method.

NOTES

[1] N. Hartmann, in his *Grundzüge der Metaphysik des Erkennens*, 2nd edition, Berlin 1925, has recently defended the opposite thesis, that we have to assume transintelligible items that are in principle inaccessible to cognition. A discussion of his arguments for this new thesis of his will be undertaken in a different setting.

[2] 'Mathematische Existenz', *Jahrbuch für Philosophie und phänomenologische Forschung* **8** (1927), 439–809, p. 502. In his earlier 'Beiträge zur phänomenologischen Begründung der Geometrie und ihrer physikalischen Anwendungen', ibid. **6** (1923), 385–560, Becker called this assertion the 'principle of transcendental idealism' (p. 387f).

[3] It is however largely independent of basic philosophic views. For whether non-empirical all-statements and existential statements are regarded as *a priori* judgments or as disguised tautologies alters nothing in their basic difference from the empirical kind of such statements.

[4] However, one important difference with regard to Kant's *a priori* is that the latter concerns only the foundations of experience as such but not of particular factual experiences, such as for example the nature of colour.

[5] In recognizing that a grasp of being-so for an arbitrary examplar constitutes a specific mode of cognition, we furthermore dispose of the sensualist theory of abstraction, according to which abstraction consists in gathering common features of empirically given objects. As Husserl has shown in his second Investigation *Logische Untersuchungen*, 3rd edition, Halle 1922, II/1 (p. 106ff.) [E.T. by J. N. Findlay, *Logical Investigations*, London 1970, p. 337], this theory contains a vicious circle, since it presupposes that the shared features to be isolated are already determinate. This erroneous presupposition then leads to the consequence that every property unambiguously determines a totality of things of that property (principle of comprehension).

Russell recognized that this principle of abstraction is untenable but failed to give up the principle of comprehension, as consistency requires.

[6] Husserl, *ibid.*, 3rd Investigation.

[7] *Ibid.*, p. 261. [E.T. p. 463]

[8] *Ibid.*, p. 266f. [E.T. p. 467f.]

[9] *Ibid.*, p. 267. [E.T. p. 468]

[10] *Ibid.*, p. 268. [E.T. p. 469]

[11] *Ibid.*, p. 276. [E.T. p. 479]

[12] Let us briefly mention the main metaphysical doctrines that have their rational origin at this point:

A realism of concepts arises as soon as qualities, their aspects and so on, are viewed as something existing independently; if these concepts or the existences intended by them are moreover regarded as value carriers, we reach the 'realm of ideas'.

The 'thing in itself' supervenes if we make the objective components of perception independent; the basis for this lies in the mental isolating process which is rooted in the fact that the content of perception can vary in two directions, as we have observed.

The assumption of a domain of incorporeal spirits or a domain of an objective spirit will arise if we make the mental independent, that is if we lift the mental out of its fundamental connection in the concrete psychophysical thing. A closely related epistemological assumption is that of independent teleological causes, which has always played an important role in metaphysics. An additional feature here is the mixing up of 'purpose' as an intentional state of affairs (idea of purpose) and as a real state of affairs.

[13] *Ideen zu einer reinen Phänomenologie und phänomenologischen Philosophie*, Halle 1913, p. 25. [E.T. p. 71].

[14] Cf. *ibid.*, p. 26, § 13 [E.T. p. 72]: "We must sharply distinguish between the circumstances of generalization and specialization on the one hand, and the essentially different ones of generalization of factual material into purely logical form, or conversely the turning into facts of logical formalities on the other hand. In other words: generalization is something quite different from formalization, for example of the kind that plays

so vital a role in mathematical analysis; while specialization is something quite different from deformalization, such as 'filling in' of an empty mathematical form or of a logical truth."

[15] Husserl himself, however, denotes not 'colour' but 'sense quality' as highest kind (*ibid.*, p. 25 [E.T. p. 71]); but this seems to me mistaken in view of his general observations about the concept of 'highest kind'.

[16] More accurately: about red or coloured objects as such.

[17] Cf. Hilbert 'Über das Unendliche', *Math. Ann.* **95** (1925), 161–190, p. 173. "In general, from the finite point of view, an existential statement of the form 'there is a number with such and such a property' has a sense only as a partial statement; that is, as part of a more fully determinate statement, whose precise content, however, is irrelevant for many applications." Our view thus differs from Hilbert's only in that we do not regard finitism as one 'standpoint' along which other standpoints might exist, but as an irresistible dictate of reason. Hilbert's thesis just quoted largely agrees with the view of Brouwer; but there remain considerable differences of opinion about the consequences, as we shall see.

[18] This applies also to the question, where 'one and the same' thing that is at place P_1 at time t_1 will be at time t_2. For in this identification determinations as to being-so are already included.

[19] We cannot here discuss the criteria of verification.

[20] *Logische Untersuchungen*, vol. 2, p. 110f. [E.T. p. 340f.].

[21] Of course it may be that expressions gain their full sense only when we take into account the personal or local and temporal data of the fact expressed. This is especially so where personal and demonstrative pronouns or spatial and temporal adverbs figure in the propositions (for example 'you have insulted me', 'the cross-roads is not here'). In such cases Husserl speaks of 'essentially occasional expressions', where it is essential "to orientate their present import according to the occasion, the person speaking and the situation" (*ibid.*, II/1, p. 81. [E.T. p. 315]) In order to decide the truth or falsity of such propositions, we must first eliminate the occasional expressions and replace them by corresponding objective determinations.

[22] Cf. Husserl's detailed analyses in his new work 'Formale und transzendentale Logik' (*Jahrbuch für Philosophie und phänomenologische Forschung* **10** (Halle 1929), 1–298, esp. p. 93ff.) about the connection of 'formal apophantics' and 'formal ontology'.

[23] *Logische Untersuchungen*, II/1, p. 46 [E.T. p. 286].

[24] l.c., p. 47 [E.T. p. 287].

[25] Cf. Wittgenstein, *Tractatus Logico-Philosophicus*, with an introduction by Bertrand Russell, 1922 London (English–German); p. 67, 4.024: "To understand a proposition means to know what is the case, if it is true." This highly important book of Wittgenstein's, to which we shall have occasion to refer more than once, had already appeared in *Annalen der Natur- und Kulturphilosophie* **14** (1921), but the London edition shows several corrections [The 1971 edition of a later translation from the same publishing house is used in this book.] A detailed account and analysis of Wittgenstein's doctrine will be given in a book by F. Waismann, *Logik, Sprache, Philosophie*, to be published shortly in the collection *Schriften zur wissenschaftlichen Weltauffassung*, Springer, Vienna. [The final version of this book, incorporating later views of Wittgenstein's, was posthumously published in English as *Principles of Linguistic Philosophy*, Macmillan,

London 1965 and in German, under the original title by Reclam, Stuttgart 1976.]

[26] Cf. Section II, note 16.

[27] The extensionality thesis goes back to Wittgenstein and, following him, was maintained above all by F. P. Ramsey, 'The Foundations of Mathematics', *Proceedings of the London Math. Soc.* (2) **25** (1927), 338–384, [*The Foundations of Mathematics*, London 1961, pp. 1–61.] and by R. Carnap, *Der logische Aufbau der Welt*, 1928, Berlin-Schlachtensee. Carnap proves the thesis l.c., p. 57ff. [E.T. p. 93ff.] by emphasizing the difference between 'sense statements' and 'reference statements', using the terminology of Frege ('Über Sinn und Bedeutung', *Zeitschrift für Philosophie und philosophische Kritik* **100** (1892), 25–50 [E.T. pp. 56–78]).

[28] Concerning the difference between 'meaning' and 'imagining' cf. the fundamental explanations of Husserl in *Logische Untersuchungen*, vol. 2, p. 61ff. [E.T. p. 291ff.].

[29] Cf. Section III below.

[30] For symbolic logic, whose main goal lies in banishing mistakes of logical thought by improvements of the symbolism, the identification of logic and language will readily suggest itself. Since, however, the symbolic logician is nevertheless not inclined to regard the principles of logic as variable, as the rules of existing symbolisms certainly are, he is often led to understand by 'language' a formally perfect language. That is the meaning of the term 'language' for example in Wittgenstein. Statements about such an ideal language are therefore only transformations of statements about the structure of the world.

[31] That is, concepts of empirical objects.

[32] Cf. below p. 31ff.

[33] Above all as regards the "pertinent questions whether the field of possibility is greater than the field containing everything actual and this in turn greater than the aggregate of what is necessary". (Kant, *Critique of Pure Reason*, 2nd edition, Transcendental logic, System of the principles of pure reason, refutation of idealism.)

[34] It would be wrong to object that a definition could never be incorrect, on the grounds that we are free to give whatever sense we wish to a term not hitherto used. For the incorrectness lies precisely in not assigning to a term the sense that belongs to it in actual use. If therefore we assert that the definition just mentioned for logical denotation is incorrect, we wish to express by this that it does not capture the sense in fact associated with it in the statements of logic.

[35] This observation is important especially for assessing Brouwer's theses on the theory of mathematics. Cf. below p. 52ff.

[36] A relation of logical denotation to numerical quantities is present only insofar as in no domain can more objects fall under a concept of smaller denotation than under one of larger denotation, although there need not be more individual objects falling under the latter than under the former.

[37] We are here thinking only of general concepts. Whether it is advisable to speak of individual concepts at all cannot be discussed here.

[38] Cf. Wittgenstein, l.c., 1.11 and 1.12: "The world is determined by the facts and by their being all the facts. For the totality of facts determines both what is the case, and also whatever is not the case." The finite formulation in the text is to emphasize that we cannot speak of the world as a closed totality.

[39] The observations just made as regards the sense of negation and conjunction in

empirical statements can be adapted to *a priori* statements simply by putting any arbitrary domain in place of some particular one.

[40] Here and in what follows we take the words 'assert' and 'assertion' as covering not only affirmation but also negation.

[41] Cf. especially Wittgenstein, l.c., 4.46ff.

[42] H. M. Sheffer, 'A set of Five Independent Postulates for Boolean Algebras, with Application to Logical Constants', *Transact. Amer. Math. Soc.* **14** (1913), 481–488. Also J. G. P. Nicod, 'A Reduction in the Number of the Primitive Propositions of Logic', *Proc. Cambridge Philos. Soc.* **19** (1917–20), 32–41.

[43] The deep and important investigations as to how an object is grasped as identical in different acts of thinking aimed at it belong elsewhere. Cf. particularly Husserl, 'Formale und transzendentale Logik'.

[44] It is to Wittgenstein that we owe clear insight as to the sense of 'identity' (l.c., 5.4733 and 5.53–5.5352). He links this with a criticism of the relevant symbolism in *Principia Mathematica* (cf. below p. 45), which Russell in principle accepts, as is clear from his preface to the *Tractatus*. This is important for what follows because the attempt, in *Principia Mathematica*, to establish Cantor's concept of the power set (Cf. below p. 122ff.) as logically legitimate makes essential use of this incorrect interpretation of identity (vol. II, p. 458ff.) Incidentally, in the definition of the individual natural numbers in *P.M.*, identity is also used in an essential way. For example, its definition of the number 1, if translated into ordinary words, is as follows: 1 is the class of all classes α for which there is an x such that x is an element of α and for every y that is an element of α, $x = y$ holds.

[45] Cf. Husserl, 'Formale und transzendentale Logik', p. 197ff.

II

SYMBOLISM AND AXIOMATICS

A comprehensive grasp of the character of the symbolisms of phonetics and writing is inseparable from a general insight into the nature of symptoms.

A state of affairs S_1 is called a symptom for a state of affairs S_2 if from the existence of S_1 we can draw inferences regarding the past, present or future existence of S_2. That such inferences can be made evidently means simply that there is a real relation (an empirical connection) between S_1 and S_2. As this definition shows, this real relation need not be such that the symptom (cognitive ground) for a state of affairs coincides with one of its causes (real ground): the symptom might equally be an effect of this state of affairs or share some causes with it. Yet even where the symptom is a cause the position is not that an especially useful symptom has to be an especially important[1] cause. Lack of clarity about the relation between symptom and real ground has caused much confusion in philosophy.

Symptoms in this general sense may alternatively be called indications, following Husserl.[2] Amongst these we must emphasize signs, used by rational beings to communicate with each other. The most important such specification of signs is language.

The words and sentences of language express contents of awareness; that is to say, the speaker, by means of certain acoustic phenomena that he produces, conveys (communicates) the content of his conscious acts to the person addressed. Expressions are therefore symptoms for contents of consciousness that can be inferred from them, and the meaning of the expressions are these inferrable contents. If the person addressed is actually to understand what the speaker means by his phonetic signs, he must 'have command of the speaker's language'; that is, he must know the schema of co-ordination used by the speaker. That in any language certain sound combinations mean something in particular, or, as we say, 'have a certain meaning', does not signify that the meaning of these acoustic phenomena is an occult quality residing in them, but merely that a number of people, as 'members of the same linguistic community', uniformly co-ordinate these sound combinations with certain contents of consciousness, so that for these signs there exists a domain of mutual understanding. To translate a word from one language into another therefore means: to ascertain that sound combination S_2 which in the human community H_2 is used as a symptom for a content of consciousness C to which in community H_1 the corresponding symptom is the combination S_1.

A language is thus the set of co-ordinating relations holding for a certain domain of intelligibility. If therefore language is simply the 'mode of expression' used by everybody within a certain domain, nevertheless to the individual, to whom it is 'given in advance' by education and perhaps in some measure even by heredity, to the individual who 'grows into' it and as a rule cannot transform it[3], at least not significantly: to him it will seem something objective existing quite independently of human subjects. Since the various creators and their shares in the created structure are usually unknown, because the creation of language rarely proceeds in full awareness and finally because this 'creation' usually occurs through a gradual process extending over many generations, some enquirers pretend to see in language nothing in the least created, but something that has 'grown'. In emphasizing that anonymous relations between individuals should not be reinterpreted as being above individuals like some 'objective spirit', we do not deny that there are important differences between popular and literary language, although the latter lacks the former's 'colour and tang' (or roots in the emotions) and wealth of untapped cognitive resources. Even if such reinterpretations sometimes afford incentives to specialist enquiry, in the end they are a serious danger to it because they stand in the way of exact analysis of the object of cognition.

These observations are meant to highlight two things above all: the conventional character of language and the relation of language to thinking (or to the grasping of what is). By an understanding of this relation, however, we do not mean the clarification of the question how far thinking actually does or can occur without one's imagining linguistic signs, but rather we are emphasizing the subsidiary character of language as a form of expression vis-à-vis thinking. Language as such, within the framework of its expressive function, can never enrich or transform being, but can only 'picture' it; leaving aside the state of knowledge at the time in question, the kind of 'picturing' will be determined in large measure by what purposes a language happens mainly to serve. Accordingly, everyday language will generally not represent the structural relations of being with the precision that is required for complex logical investigations. From the point of view of practical epistemology it is therefore a very important achievement that in recent decades symbolic logic has evolved as a system of signs especially designed for the purpose of investigating the foundations of mathematics. The fact that in this language the visual aspect for technical reasons displaces the acoustic, is in principle irrelevant.

Before turning to the logicist symbolism, let us be quite clear what is meant when we speak of the objective meaning of language or of certain linguistic signs. For there is here an ambiguity of conception that is apt to create confusion.

On the one hand by the 'objective meaning' of linguistic signs we mean those thoughts that in a certain community are generally expressed by these signs. Here 'objectivity' means intersubjective thinking within such a domain, without anything being stated about the content of thoughts.

More narrowly, however, linguistic signs are said to have a 'sense' or to be 'objectively meaningful' if they express meaningful thoughts: here the 'objectivity of meaning' lies in the fact that thoughts relate to 'objective being', to objects and states of affairs in the world.

This must be correctly understood: as we have observed in the first section, all thinking is of something assumed to be independent of the fact that it is thought, which excludes there being a kind of thinking completely divorced from any link with being; but it is possible that 'connections' in thought may gain a footing to which no connections in being can correspond.

Take for example on the one hand the concepts 'silly circle' or 'four-cornered virtue', and on the other 'black piebald' or 'young crone'. Both varieties share the assembling of what does not belong together, but in the first group this lack of belonging is more radical. For there, to the concepts combined correspond highest kinds that are themselves incompatible, while this is not so for the second group. It is important to draw a sharp distinction between these two cases, as Husserl has done with his distinction between 'nonsense' and 'counter-sense'[4]. In particular, we must emphasize that the antinomies of logic and set theory largely rest on 'nonsensical' presuppositions. Such nonsensical combinations stand beyond the applicability of the principle of contradiction, for that principle presupposes meaningful statements, as we saw in the previous section.

In this connection it should be pointed out that the ascertaining or negating of 'nonsense' or 'absurdity' is based exclusively on a priori knowledge and therefore does not presuppose any enquiries as to empirical existence. Following Leibniz, this is often expressed by saying that these relations hold 'for all possible worlds'; but this way of speaking does not seem to me felicitous, since closer analysis shows that the concept of a 'possible world' simply means the domain of validity of truths of reason, that is judgments a priori. If, then, a priori and a posteriori knowledge relate to one and the same world as their domain of validity, still we must not blur the difference in principle between these two ways of knowing.

For what follows, the two most important results of this investigation are:

(1) The concept 'senseless sign' is a contradiction in terms, for the assertion that visual or acoustic phenomena are 'signs' already contains the assertion that by means of these signs one can understand something, that with their help one can grasp the thoughts of others.

Of course it need not be the case that each spatially or temporally independent visual or acoustic phenomenon has an independent sense; rather, it may be that sense does not accrue until several such phenomena have been connected in certain ways. It is not entirely correct to speak of 'dependent signs' in this case, since no sign at all is present until such a connection, and with it a meaning relation, is produced.[5]

(2) Through its sense, that is through its connection with thinking, a sign is indirectly linked with the existent that forms the object of thinking. Accordingly, it is essential for every sign that it should mean something that is, namely that which forms the object of the thought it expresses.[6]

However, these two observations seem to be inconsistent with what is perhaps the most important new finding in the investigation of the foundations of mathematics, namely Hilbert's theory of proof.

Let us convey its basic ideas in Hilbert's own words: "The basic idea of my theory of proof is simply to describe the activity of our understanding, to place on record the rules according to which our thinking actually proceeds."[7]

"The statements that make up mathematics are all transposed into formulae, so that mathematics proper becomes a stock of formulae. These differ from ordinary formulae of mathematics only in that they contain, besides the usual signs, the additional logical signs $\rightarrow$, 'follows'; &, 'and'; V, 'or'; —, 'not'; (x), 'all'; (Ex), 'there is'. Certain formulae, which serve as building bricks for the formal structure of mathematics, are called axioms. A proof is a figure that must be intuitively present to us; it consists of inferences according to the inference schema S, $S \rightarrow T$, T, where the premisses (that is the formulae S and $S \rightarrow T$) are always either axioms or arise directly by substituting from an axiom or coincide with the final formula for an inference previously occurring in the proof or arising from it by substitution. A formula is to be called provable if it is either an axiom or the end formula of a proof.

"Axioms and provable propositions, that is formulae that arise from this procedure, are pictures of the thoughts that make up the whole of customary mathematics."[8]

"Mathematics proper thus formalized is then supplemented as it were by a new mathematics, a metamathematics which is needed to secure the former, in which, by contrast with the purely formal modes of inferring in mathematics proper, the inference concerns content, but only in order to show that the axioms are consistent. In this metamathematics we operate with the proofs of mathematics proper, these latter themselves forming the content investigated."[9]

"This problem of consistency, however, is entirely accessible to treatment given the present position of things. As is immediately obvious, the problem amounts to recognizing that we cannot by means of the rules stipulated, deduce from our axioms the end formula $1 \neq 1$, which is therefore not a provable formula."[10]

The view expressed in these quotations needs critical correction in one particular; but this observation is by no means directed against the theory of proof as such, whose mathematical and epistemological scope is in my opinion very great indeed, as will no doubt become increasingly clear in years to come: our remarks are directed only against the philosophical interpretation of this theory by Hilbert and Bernays, its creators.[11]

On this interpretation, which emerges from the above quotations, proofs are intuitively given figures quite devoid of meaning. Against this our analysis in Section I (p. 33f.) has made it clear that the rules for the use of signs in the formulation of 'figures of proof' themselves contain the sense that belongs to logical transformations as such, (belongs, that is to the formation of statements whose meaning is contained in that of the statements already given). This is brought out especially clearly by the fact that in the proof figures too we use different groups of signs, which on interpretation of content enable us to differentiate between individual signs and signs for variables, as well as between formally different kinds of individual and variable signs.[12]

If therefore the signs and formulae even of Hilbert's theory of proof have meaning (and his is the most radical formalisation conceivable), this will hold a fortiori for the implicit definitions his geometrical axiom system rests on.

Hilbert's axioms of geometry are in fact statements about certain arithmetical, or in the narrower sense logical, relations between arbitrary objects; that is, a logico-arithmetical (relational) schema that can be variously filled in by intuitive or pseudo-intuitive objects.[13] That in his axiomatics for geometry there are exactly three systems of objects and that between them numerical and other ordering relations are stipulated, this alone is enough to constitute the meaning content of this axiomatic system. The implicit definitions of objects therefore amount to the stipulation of formal relations between otherwise arbitrary objects.

From this we can further explain how geometrical relations can be represented on arithmetical relations, a fact that Hilbert uses to prove that his geometrical axioms are mutually independent and consistent. For this possibility rests on the circumstance that the so-called geometrical relations of Hilbert's axiomatics, in which all reference to intuition is eliminated, are in fact logico-arithmetical relations; the co-ordination merely brings out the character of these relations more clearly. We shall deal with this point in more detail in Section IV.

Finally this sheds light on the problems of isomorphism. By this we understand the possibility of a one-one representation of two axiomatically stipulated domains in such a way that the holding of arbitrary relations between the elements of the one domain implies the holding of certain co-ordinated relations between the corresponding elements of the other domain. This amenability to representation is often accepted as an ultimate datum that cannot be further explained. However, here as in most other cases, it is a sign of insufficient insight if we operate with definitions of possibility; for possibility now points with especial force to an existential basis or criterion that 'generates' it, and the determination of that criterion becomes the real problem: thus for example with 'constructibility', 'possibility of well-ordering', 'decidability', as we shall show later. Likewise with isomorphism, which is simply a formal (structural) sameness, while the indication of a correlation that establishes a one-one co-ordination between the isomorphic relations is merely a means for coping with this formal sameness. It is a means (and not the only one either), since the structure of each system exists independently of that of any other system; therefore there must be a logical 'normal form' that enables us directly to test any two axiom systems as to isomorphism. The theory recently worked out by Hilbert and his collaborators seems to be the appropriate instrument for this.[14]

In the carrying out of any such investigations, symbolic logic plays an important part so that we must be clear on which elements in it make it suitable for the purpose in hand.[15]

First we observe that symbolic logic is not designed to represent any differences other than structural ones; from the outset it confines itself to describing the structure of the world.

There is the further question on what principles a functional expression must be based in order that the logico-mathematical operations that have to be carried out should be most convincingly illustrated by the signs employed. Symbolic logic tries to achieve this by choosing the signs in such a way that the structure to be expressed is pictured in them, or, we might say, 'shows' itself in them.[16]

Accordingly, objects (carriers of relations) that are assumed to be alike will be co-ordinated with the same sign; and if different, with different signs.

The representation of a relation between a finite plurality of different things is then performed in such a way that to every object one sign is assigned in mutual one-one co-ordination. If a relation is composed of several others, this is reflected in that the notation for it will be likewise composed of the signs for the simpler relations.

This greatly enhances logical perspicuity and brings every single step of the thinking process into full consciousness, which makes mistakes much more easily detectible.

However, every 'language' has certain irremovable defects because it must order the individual acoustic or visual signs in space or time by setting them after or alongside each other, or above or below, which introduces into the picture a 'direction' that the relation to be represented does not itself possess.[17]

Let us illustrate this by a simple example of ordinary language. The statement 'the male person A is a brother of the male person B' expresses the same state of affairs as the statement 'the male person B is a brother of the male person A', although the judgments differ in meaning, since the first has A as subject and the second B. This fact must not be interpreted as though the two statements 'first' relate to two separate states of affairs whose identity is ascertained only afterwards; but rather there is only one state of affairs and the doubling up arises merely from the way of denoting it which imports a direction into the essentially non-directional relation by the terms of the relation having a different position in space or time.

For what follows, this yields the very important observation that a symbolic description of relations must not be viewed as though the distinctions thus made (such as transitivity, non-transitivity, intransitivity) revealed properties of the various relations, which would amount to saying something about the world. For in fact this description does not touch the relations themselves but the different ways in which they are symbolized and their mutual substitutability. We must therefore emphasize that in contrast with prevailing opinion there is here no cause for setting up a calculus for symbolizing the properties of relations. That such an 'extended functional calculus' is in any case not required follows from the analyses in Section I (p. 23f., 28ff.).

We must point once more to a fact emphasized in Section I: the way we denote situations puts a direction in place of what is as such non-directional; yet this agrees with the direction of thinking, inasmuch as thinking brings a mental order into individual cognitive contents and their relations, by means of the temporal succession of their apprehension. Let us illustrate this by the example of implication. '*A* implies *B*' means '*A* and non-*B* never exist together', a relation in which *A* is in no way prior to non-*B*. Thinking, however, starts from the existence of *A* and then infers from it the existence of *B*, which may well not be immediately obvious; so that *A*, as being apprehended first, figures as prior to the inferred *B*, and the order of thought is then wrongly interpreted as an order of being. This misinterpretation has greatly contributed to obscuring the basic problems of logic and mathematics.

We now give a brief and incomplete summary of the concepts essential for the description of the above relations, as found in Carnap.[18]

"A relation is called symmetrical if it is identical with its converse (e.g. being the same age as); otherwise it is called non-symmetrical (e.g. brother); a non-symmetrical relation is called asymmetrical if it excludes its converse (e.g. father). A relation is called reflexive if, in the case of identity (within its field), it is always fulfilled (e.g. being the same age as); otherwise it is called non-reflexive (e.g. teacher). A non-reflexive relation is called irreflexive if it excludes identity (e.g. father). A relation is called transitive if it always holds for the term next but one as well (e.g. ancestor); otherwise, non-transitive (e.g. friend). A non-transitive relation is called intransitive if it never holds for the term next but one (e.g. father). A relation is called connected if, between any two different terms of the field, either it or its converse always holds (e.g. for six people seated round a table, the relation 'one, two, or three seats to the left of'). A relation is called serial if it is irreflexive and transitive (and hence asymmetrical) and connected (e.g. 'smaller than' for real numbers). A relation is called a similarity if it is symmetrical and reflexive, and an equivalence if it is also transitive."

We must now focus our attention on a distinction not mentioned in this list, namely that between one-one, one-many, many-one[19] and many-many relations. That a relation between two objects is one-many or many-one means that one of the two objects may stand to other objects in the same relation as to the second of the two, while the second cannot share its relation to the first with any other objects. However – especially in the sphere of being-so, to which we here confine ourselves – relations between a particular object and two or more mutually different objects cannot be quite the same, but only up to a point, which leaves some room for differences. Therefore what corresponds to one-many or many-one relations (and a fortiori to many-many relations) are not peculiar states of affairs not encompassed by one-one relations, but they are merely logical combinations of one-one relations between eidetic singularities. For we must realize that all relations in the domain of being-so are built up from relations between eidetic singularities. A complete determination of the relations between the eidetic singularities in a given factual or formal sphere determines all relations in that sphere.[20]

If on the one hand the symbolic logical calculus, like mathematical symbolism in general, serves to represent, in as unadulterated a form as possible, the formal layer of being with which logic and number theory are concerned, on the other there arises a danger of forms of logical and mathematical symbolism becoming so intimately fused in thought with the logico-mathematical objects symbolized, that these latter are no longer clearly distinguished from the ways of symbolizing them. As a result, in mathematics, the introduction of new symbols for heuristic purposes is regarded as a 'free creation of the spirit' and therefore as a real extension of the sphere of mathematical objects.[21]

In view of this, let us clarify the fundamental difference between mathematical objects on one side and the way they are determined or represented on the other, in terms of a particularly impressive example, namely the decimal notation for natural numbers. Here the difference lies so close to the surface that no observant critic could miss it; but in principle the position here is no different from that attending infinitesimal symbolism or ideals in number theory.

Our example concerns the representation of natural numbers in the decimal system; that is, as series of powers with the base 10. As regards the privileged position of the number 10, which is doubtless linked with the not purely mathematical fact that a man has ten fingers, every mathematically more or less educated person will know that the cognitive content of arithmetic changes not at all if, instead of 10, we take any other natural number greater than 1 as basis. Yet representing natural numbers by series of powers, however suitable it may be, is by no means a method essential to arithmetical thinking; any state of affairs concerning natural numbers can indeed be formulated without this mode of representation, or transformed into another mode of representation.

Both logical and mathematical procedures consist in such transformations; their significance can be correctly grasped only if we are perfectly clear on the distinction between mathematical objects and the way they are represented.

If, however, one denies that mathematical objects are independent of symbolism, one will be inclined to over-rate a change (such as an enrichment of the symbolism). This is manifest in Hilbert's method of ideals. The introduction of ideal elements serves to simplify mathematical statements and proofs. As examples of this method, Hilbert mentions the formation of algebraic symbols, negative and complex numbers and Kummer's ideals, and then aims to view both the denumerable and the non-denumerable infinite as likewise an aggregate of ideal elements.

"Just as $i = \sqrt{-1}$ was introduced in order to maintain the laws of algebra in their simplest forms (for example, the laws concerning the existence and number of roots of an equation); just as ideal factors were introduced in order to retain for algebraic integers the same laws of divisibility (for example, by introducing an ideal common factor for 2 and $1 + \sqrt{-5}$ although there is no real one); so we must adjoin ideal statements to finite statements, in order to maintain the formally simple rules of customary Aristotelian logic. It is strange that the modes of inference so passionately attacked by Kronecker are the exact counterpart of what the same Kronecker so enthusiastically admires in Kummer and number theory, praising it as the highest mathematical achievement."[22]

On Hilbert's view, the only barriers of principle to introducing ideal elements are that the relations (operations) defined for these elements must fit consistently into the system of mathematics, so that their 'construction' (reduction to the basic elements of mathematics) is not required; accordingly, for Hilbert, statements about ideal elements are pure existential statements, and his way of justifying these on the basis of their heuristic importance entirely agrees with what he does with regard to the introduction of ideal elements. For example, in one of his most recent lectures on foundational questions he says: "what is valuable in pure existential proofs is precisely that by means of them individual constructions are eliminated and many different constructions are gathered into a basic thought, so that what stands out clearly is only what is essential for the proof: the point of existential proofs is abbreviation and economy of thought."[23]

From these and other places in Hilbert's more recent programmatic explanations concerning his theory of proof[24] it seems to follow without any doubt that the only (and indeed only possible) aim of introducing ideal elements into mathematics and of operating with existential statements is to simplify formulation and demonstration; of course, this will very often be decisive for the success of heuristic efforts and mathematics can hardly be conceived as doing without it. However, the introduction of ideal elements does not create new mathematical objects.

Now Hilbert tries to use the method of ideal elements to secure for the non-denumerable infinite of Cantor's set theory a legitimate place in mathematics, although it is in principle impossible to construct it, as we shall see in what follows.

For what is involved here, even if perhaps not with full awareness, is the sham 'existence' based on the principle of comprehension (according to which any mathematical property determines a totality of objects having that property). Assertions about such 'existence' are, however, in principle different from those about the existence of finite bounds[25]: our objections, unlike Brouwer's, are aimed only at the former.

If, accordingly, Hilbert has given up an 'actual transfinite' while nevertheless believing that he can maintain it in 'formal mathematics', since here the only restriction on thought is contradiction, we must reply that the uniqueness of this barrier for the domain of mathematical thinking remains incontestable, but that 'sense' (in mathematical thought: reference to the formal sphere) is the prerequisite for entering the domain encompassed by the barrier of contradiction. If, however, even finite figures of proof are regarded as 'senseless', the genuine division between sense and absence of sense remains unnoticed.

While fully grasping how important, indeed in practice indispensable for cognition, mathematical symbolism in its present form is, we must not give up the observation that every proposition formulated with the help of that symbolism must retain sense when the symbolism is dissolved except for certain basic symbols; and that if a proposition set up with the help of certain symbols is a tautology or a contradiction, it remains so after 'translation' into a language not using those symbols. This observation can be formulated as the principle of the independence of mathematics from mathematical language, which L. E. J. Brouwer has recently placed at the head of his criticism of Hilbert's formalism'.[26]

This thesis is closely connected with that of the constructibility of all purely mathematical objects by means of the natural numbers (which corresponds to Brouwer's earlier formulation), if by construction we understand 'definition'.[27] These theses of Brouwer's must be entirely accepted.

However, near to the Scylla of fusing mathematical symbolism with mathematical objects leading to the 'extravagant' use of symbolism and thence to formation of pseudo-objects, there lies the Charybdis of merging mathematical objects with the process of mathematical thinking which runs in time and this amounts to regarding mathematics as dependent on the momentary fact of mathematical cognition.

This temptation is particularly attractive for those who wish to grasp the nature of mathematics by starting from an analysis of thinking, and that for two reasons: firstly, because the foundation of mathematics, namely number, is closely connected with the temporal process of counting, though, as we shall see in the next section, time does not enter into number; and secondly, because a totality of all numbers does not really exist so that there seems to be no fixed domain of mathematical cognition given before the event.

Accordingly, says Brouwer, we do not adequately grasp the sense of mathematics if we regard it as something that has being; rather, it is a becoming, a mental act. His doctrine, which he calls neo-intuitionism,[28] has strongly influenced H. Weyl[29] and O. Becker.[30] Let us proceed to describe briefly the basic conception of Brouwer's intuitionism, which is at the centre of current controversy concerning mathematical method; following this we shall give a short critique of its principles.

In order to do justice to Brouwer's doctrine we must keep in mind that it sprang mainly from an endeavour to resolve the absurdities connected with the non-denumerable infinite in mathematics by means of a radical analysis of mathematical thinking. Only thus can one find a correct approach to Brouwer's theses.

Brouwer clearly recognizes that it is wrong to assume the existence of infinite totalities and therefore states with all desirable candour that an infinite set is nothing but a law; nevertheless, his other main and closely interrelated tenets, namely denial that the principle of the excluded middle is universally valid, denial that arithmetic and analysis are definitely decidable, exclusion of pure existential propositions and existential proofs, all these are preponderantly directed towards the problems of the non-denumerable infinite, which are really settled by his above declaration.

The central point of Brouwer's attack on the doctrine of the non-denumerable infinite lies in his remark "that the axiom of comprehension, on the basis of which all things possessing a certain property are united into a set . . . is inadmissible, that is unusable, for establishing set theory, and that it is necessary to base mathematics on a constructive definition of set".[31]

Once eliminate the axiom of comprehension and it emerges clearly that the whole of mathematics can be reduced to the natural numbers Brouwer puts it thus: ". . . all mathematical sets of units which are entitled to that name can be developed out of the basal intuition, and this can only be done by combining a finite number of times the two operations 'to create a finite ordinal number', and 'to create the infinite ordinal number ω'; here it is to be understood that for the latter purpose any previously constructed set or any previously performed constructive operation may be taken as a unit. Consequently the intuitionist recognizes only the existence of denumerable sets, i.e., sets whose elements may be brought into one-to-one correspondence either with the elements of a finite ordinal number or with those of the infinite ordinal number ω".[32]

How, then, does Brouwer view as given the infinite ordinal number ω, which in Cantor's set theory figures as the ordinal number of the well-ordered set of all natural numbers? What sense does the denumerably infinite have for him? Here one of the roles that time, on Brouwer's view, plays in mathematics comes clearly to the fore: for he regards the denumerable as given in the form not of a sequence that is, but of one that becomes. The basic schema of such a sequence is the sequence of natural numbers. This sequence is law-like, for any term in it 'determines' a successor according to a general law, so that it is possible to symbolize this law by a general term (n) which is stated to follow that law. Brouwer next contrasts these law-like sequences with 'freely becoming sequences of choice', mainly in order to introduce an intuitionist concept of the continuum. These sequences of choice are completely free if individual numbers are successively chosen at random; besides these he recognizes sequence of choice that are free within a certain range, being subject to certain restrictive conditions. Finally, he speaks of sequences of choice whose terms are fixed by choices performed in other ways.[33]

An example of the genesis of sequences of choice of the first kind is the arbitrary putting of numbers one after another; of the second kind, the throwing of ordinary dice, the range of variation comprising the numbers 1 to 6; and of the third kind the sum of two sequences of choice of the first or second kind, such as the sum of successive throws of a die. It is obvious that in a freely becoming sequence of choice, its definite character beyond what has been realized can never reach further than the law to which it is subject; but since there cannot here be a law-like character throughout, the definite character depends at any stage on the point reached. Thus, according to Brouwer, with respect to that point there are decidable and undecidable questions. For example, if in drawing numbers from an urn we have twice drawn a 3 once a 2 and we form the corresponding number series, it remains for the time being undecided whether the series will contain a 1. There are therefore two kinds of question here: those that 'can already' and those that 'cannot yet' be decided; which is not in the least strange, since what we are concerned with is just to characterize empirical objects, where the difference between being definite and indefinite coincides with that between having become and still becoming. So far so good.

However, Brouwer proceeds to apply the reflections just outlined to analysing mathematical problems by asserting that as regards the decidability of mathematical problems for which no method of solution is known, cognition is in the same position as towards free sequences of choice; the answer is neither 'yes' nor 'no', since the state of affairs in question remains currently indefinite. On his view, therefore, the principle of excluded middle does not apply here. It does hold throughout for finite matters, but not for transfinite ones.[34]

Thus his two theses (that the principle of excluded middle does not hold throughout and that mathematical questions are not decidable throughout) are equivalent as he expressly emphasizes. Whether a given mathematical problem is solvable can then be decided only by actually indicating the solution or by exhibiting the problem as a special case of a more general problem that has already been solved. For Brouwer the domain of the unprovable in mathematics is by no means fixed once and for all; rather, it shrinks as mathematical knowledge progresses, in that propositions that previously belonged to it are now ranged in the group of provably true or provably false propositions.

For example, according to Brouwer, as long as we do not know of any sequence 0123456789 in the decimal expansion of π, we cannot assert that such a sequence either does occur or (exclusively) that its occurrence is absurd (contradictory). Rather, he thinks, there are three possibilities:

first, a sequence of this kind is known or a proof is known that it begins no later than at the nth place in the decimal expansion of π, in which case we are entitled to say that the sequence as stated does exist; secondly, a proof is known that the occurrence of that sequence in the decimal expansion of π is contradictory, in which case we can say of the sequence that it 'does not exist'; and thirdly, neither a proof (construction) for nor against the existence of such a sequence is known, in which case we can assert neither existence nor non-existence.

From the assumption of this trichotomy instead of the traditional dichotomy, it follows that from its being absurd that an existential assertion in mathematics is absurd we cannot infer the correctness of that existential assertion.[35]

Let us disregard the question whether Brouwer has given particularly felicitous expression to his own position by declaring it to be the assertion that the principle of excluded middle does not hold in infinite domains,[36] and proceed at once to examining whether his view is justified.[37]

To this end we must ask, using the above example, under what circumstances we can regard it as established as absurd to assert that it is absurd that the indicated sequence occurs in the decimal expansion of π. Evidently this is the case if and only if we have proved that at some point of the decimal expansion such as a sequence actually begins. However, according to Brouwer, this does not yet satisfy the first of the above three conditions, for if there is no upper bound to the number of places, we can never be sure that we shall actually succeed in ascertaining such a sequence.

Against this thesis of Brouwer's we may begin to argue as follows. If we have proved that the sequence in question begins at 'some arbitrary place' in the decimal expansion of π, it is certain that that place can be reached in finitely many steps from the beginning; but this insight is here the only relevant piece of mathematical knowledge. For psychological and anthropological considerations about the mathematical abilities of man, or observations about the extent of current mathematical knowledge do not belong to mathematics. Yet Brouwer would not accept this objection, since he regards mathematics as mental action, so that it might look as though the controversy about the foundations of mathematics here led to ultimate and rationally unresolvable differences of outlook.

This appearance does however vanish on closer reflection. For Brouwer's view that mathematical facts change with mathematical knowledge implies that there is something cognizable that is created only by being cognized,[38] which runs counter to the nature of cognition. For, as we saw in Section I, all cognition presupposes an object that must be thought of as existing independently of its being cognized.

This statement is not a dogmatic presupposition, but a 'result of reflection', to employ a most apposite expression recently used by Brouwer himself. That is to say: if we make ourselves thoroughly aware of the sense of thought processes in general and of mathematical enquiry in particular, one understands that the objects grasped or to be grasped by thinking processes are thought of as independent of the fact of their being thought about. We can formulate this insight by saying that mathematical propositions are discovered and not invented, provided we carefully avoid misinterpreting the concept of 'discovering' as though alongside undiscovered facts there also existed undiscoverable facts;[39] that is, in our case, mathematical facts that were in principle inaccessible to exact mathematical cognition (axiomatisation).[40]

In rejecting the philosophical interpretation that Brouwer gives to his critical theses, we have, however, not abolished the fact that at various points in mathematics one operates with the absurdity of the absurdity of the existence of finite bounds, although it has so far not been possible to find a fixed value for these upper bounds. This again produces the illusion that the absurdity of the absurdity of the existence of a finite bound and the constructibility of a finite bound are different mathematical states of affairs.

In what follows we shall try to solve this dilemma, although we admit at once that this solution will no doubt not be regarded as entirely convincing so long as its credentials have not been established in terms of the most important proofs here in question.

Our train of thought is this: if we make perfectly clear to ourselves what are the criteria for a mathematical assumption being absurd, we recognize[41] that they lie in the fact that two signs for natural numbers presupposed as synonymous turn out, at the end of a proof, not to be synonymous. Every contradiction therefore relates to certain natural numbers; that it arises at all means that it occurs at certain points and in exhibiting it we must therefore implicitly determine those points. That this determination does not always show in mathematical proofs themselves must be traced back to the abbreviations in the formation of mathematical concepts (symbolism); but a completely articulated theory of proof, as aimed at by Hilbert, would have to bring it into the open. If so, the illusion that there are pure existence proofs would vanish and Brouwer's demand for constructivity throughout would have been met; but it is precisely the fulfilment of this demand that will show that wherever the question is that of the existence of finite bounds, the only question that arises in the building up of mathematics (apart from set theory, to be analysed later), a divergence between absurdity of absurdity and mathematical existence did not even 'originally' exist, so that the result of existence proofs in classical mathematics will be perfectly justified from Brouwer's point of view as well, leaving aside differences in terminology.

However, this does not hold for the pseudo-existence proofs in the doctrine of the non-denumerable infinite, which, as we shall see, do not have a constructive basis. That in rejecting these proofs, and indeed eliminating the sphere of the non-denumerable infinite from classical set theory, we are by no means shaking the foundations of mathematics, will be shown in Section IV and V.

The most important result of Brouwer's critique of method is, in my view, the dissolution of the sham domain of the non-denumerable infinite in classical set theory; to do justice to this achievement we must stress this aspect and not his theory (influenced by Kant)[42] concerning the connection of time and number.[43]

In conclusion we can say that Brouwer's critique is valid and important, insofar as it is directed against the doctrine of actual infinity in general and the non-denumerable infinite in the sense of Cantor's set theory in particular; but we should reject that part of Brouwer's doctrine which is essentially based on the introduction of the concept of time into mathematics. As to his position regarding problems of decidability, we shall return to it in Section IV.

We now proceed to a brief analysis of the axiomatic method. This method, according to the well-known definition of Weyl,[44] consists in "collecting all the basic concepts and facts from which every concept and proposition of a science can be defined or derived respectively".

Let us describe the most important requirements usually imposed on axiom systems, confining ourselves to the axioms of formal domains (a restriction that will be important in determining the concept of completeness of an axiom system); this will, however, include all geometrical axiom systems that, like Hilbert's, leave aside the intuitive meaning of the basic concepts and therefore amount to a pure system of abstract relations.

The chief requirement is consistency. The traditional method of showing that an axiom system was consistent was to indicate a 'model' for that system thus showing that the latter is not empty, for so it would have to be if it were inconsistent.[45] In contrast, Hilbert's new investigations enable us to prove that an axiom system is consistent by analysing it internally.

Of quite a different character is the requirement that the axioms be mutually independent,[46] namely the postulate that none of the axioms be deducible from any of the others, thus avoiding over-determination. For clearly this requirement is merely a postulate of mental economy. The truth (correspondence with being) of what is said evidently does not depend on whether it is said with more or fewer words. Since an axiom's being underivable from the others is equivalent to its negation being compatible with them, we can exhibit non-derivability by indicating a model in which the other axioms are joined with one that contradicts the first. Thus the underivability of Euclid's postulate from the other axioms, such as those of Hilbert's axioms of geometry, is shown by models of non-Euclidean geometry.

Geiger (l.c., p. 25ff.) has pointed out that there is a defect of principle in these underivability proofs, which conclude from existence to possibility without encompassing the reasons for this possibility. This blemish can be removed if we internally ascertain that an axiom system is consistent, as Hilbert has shown that we can in principle. Geiger's way of striving towards the same goal we shall consider later.

The requirement that the least number of basic concepts be used has the same status as the underivability condition. Still, it may sometimes be advantageous for technical reasons to use more basic concepts than is in principle necessary, in order to circumvent otherwise unavoidable symbolic complications.[47] The introduction of a new sign will, however, be highly dangerous if one is not absolutely clear what it denotes. It is a fatal mistake to think that lack of clarity about certain cognitive facts could be rendered harmless by 'formalising' what has not been clearly grasped, by introducing a specific symbol with which one proceeds cheerfully to operate without quite knowing what it is a symbol of, that is, what state of affairs it denotes. It will be one of the main tasks of this essay to lay bare what are the consequences if we operate rashly with the concept of a set.

What has been said about limiting the number of basic concepts holds likewise for limiting the number of axioms.

In addition we must mention briefly that the same cognitive facts can be described by different axiom systems and that there is considerable freedom as regards the choice of basic concepts. The best-known example in Euclidean geometry of an axiom that can be replaced while all others are retained is the equivalence of the parallel postulate and the theorem on the sum of the angles of a triangle.

One of the most important problems in axiomatics is that of the completeness of axiom systems. Three concepts of completeness must here be distinguished.[48]

(1) Completeness as monomorphism. An axiom system is called monomorphic if two arbitrary models for this system are always isomorphous, that is, one-one representable in such a way that the existence of any relevant relation between the elements of the one system implies the existence of the corresponding relation between the corresponding elements of the other system.

(2) Completeness as non-branchability. This means that any relevant assumption compatible with the axioms excludes the corresponding compatibility of its negation.

(3) Completeness as decidability. An axiom system is called decidable if every question that falls under it can be decided.[49]

However, these definitions suffer from the fact that the criteria in question are not indicated.

The questions that really need answering are precisely these: under what condition are all models of an axiom system isomorphous? Under what circumstances does the compatibility of a relevant statement with the axiom system exclude the compatibility of its negation with the axiom system? Under what conditions is every relevant question decidable on the basis of the axioms?

As to these questions we may remark that on the one hand the 'completeness' (in each of the three senses) of a given axiom system is determined 'internally' (Wittgenstein), that is independently of the empirical course of events; but on the other hand the criterion of 'completeness' of an axiom system must evidently lie in aspects that are not in turn determined by this system. These aspects must therefore be sought in the 'edifice of the world' and more particularly in its structure, since we are considering only formal axiom systems.

Once this has been grasped, it is not difficult to understand that all three concepts of completeness (monomorphism, non-branchability, decidability) point back to the same criterion, namely the unambiguous (inaccessible to further formal specifications) determination of a formal domain.[50]

Thus in a complete axiom system relations between basic concepts are fixed in such a way that any arbitrary additional relation that is not explicitly fixed would be inconsistent with the axioms.

Take an example from the axioms of Euclidean plane geometry: a straight line shares no points with exactly one of the straight lines through a point outside it. With any other straight lines through that point it shares exactly one point. Obviously there is here no room for further completion: any new (non-superfluous) relevant specification must be inconsistent with the earlier ones.

Completeness thus means complete formal determinacy which must be ascertainable by analysis of the axiom system itself without reference to any other such systems, precisely as Hilbert has shown for consistency. We shall come to see this even better in Section III where we set up the axioms of arithmetic.

Axiomatic technique has been admirably refined in recent decades by enquirers under the intellectual leadership of Hilbert and Russell; but since it was mainly designed for heuristic purposes in mathematics, certain aspects particularly important for foundational enquiry have not received adequate attention. Above all this concerns questions linked with the cognitive content of axiomatics, that is, with the 'models' whose structure the axioms describe. The attendant dangers we have already noticed.

Geiger, in his above-mentioned book, tries to meet these problems by setting up a 'systematic axiomatics of essence'. Since he proposes to give an axiomatic account of Euclidean geometry, his basic problem is this: "Is there some essential order in the construction of the objects of Euclidean space, that can be copied by the structure of deductive mathematical theories?"[51] This question may be called the problem of the axiomatics of essence. By thus considering the domain whose structure the axiomatics is to reproduce, further leading ideas are obtained for the finding of axioms, which process is the aim of systematic axiomatics. In particular, it is to guarantee the completeness of the axiom system set up. Geiger not only states his requirements as a programme, but largely carries them through in the construction of his axiomatics.

Here, in his own words, are the three principles for discovering his axioms:[52]

"1. Principle of correspondence between construction of the world of objects and axiomatics: the number of basic concepts to be recognized in axiomatics is the same as that of basic elements and relations in the construction of the world of objects. Their number must not be artificially diminished by improper definitions.[53] Likewise the course of deduction in its systematic construction must use only such inferences and derivations

as are matched on the side of the world of objects by groundings and complications. Also, in formulating theorems and axioms we must choose a form that gives equal formal expression to states of affairs that are similarly ordered, not artificially transforming factual independence into mental dependence. Hypothetical propositions should therefore be avoided. In their place assertions of incompatibilities of relations will appear. The axiomatics of essence therefore transforms 'if *S* then *P*' into '*S* and non-*P* are incompatible', or 'it is impossible for *S* and non-*P* to exist together'.
"2. Principle of exclusion: characterizing axioms[54] within systematic axiomatics must not be regarded as stipulations of positive content, as assertions about possible existence; but as assertions about the being excluded of certain mathematical possibilities. All characterizing axioms are axioms of exclusion. Their systematically correct formulation ... therefore runs: it is impossible that this or that possibility is realized in the qualified[55] world of objects.
"3. Principle of the mathematically systematic character of axioms: the search for axioms proceeds by systematically ordering the mathematical possibilities, and every such possibility is examined as to its being realizable or not, within the qualified world of objects."

These investigations by Geiger on axiomatics are specially emphasized here because their basic conception is highly important for the theory of mathematics even beyond the framework of axiomatics. It is the 'objective' view which is aware that a radical solution of the foundational problems of mathematics can be reached only if one has clearly grasped the character of 'mathematical objects', that is, the theme of mathematical enquiry: within the sphere of foundational enquiry we should always remember this insight. This conception is one of the leading ideas of the present essay and will stand out especially in the investigations on natural numbers to follow presently.

NOTES

[1] The 'essentiality' of causes of a phenomenon is here determined by the 'extent' (inductively ascertained) of changes that are observed in the phenomena when those causes are absent, other things remaining the same. This does of course not yet define the concept precisely.

[2] *Logische Untersuchungen*, vol. 2, p. 24 [E.T. p. 688].

[3] That is, in most cases he cannot bring the members of a linguistic community to an enrichment or change of the phonetic symbolism uniformly used by them (or, if he can, then only in a very modest measure).

[4] *Logische Untersuchungen*, vol. 2, p. 326 [E.T. p. 516f.]; on this view examples of the first kind should be denoted as 'nonsense'.

[5] Irrespective of this, the term 'incomplete symbols' has gained currency in these cases. Cf. A. N. Whitehead and B. Russell, *Principia Mathematica*, Cambridge, vol. I, 1910 (2nd edition 1925), vol. II, 1912 (reprinted 1927), vol. III, 1913 (2nd edition 1927). Unless otherwise stated, quotations will be from the first edition. Vol. I, p. 69: "By an 'incomplete' symbol we mean a symbol which is not supposed to have any meaning in isolation, but is only defined in certain contexts."

[6] Cf. also the difference between 'expression' and 'meaning' in R. Carnap, *Der logische Aufbau der Welt*, p. 24f. [E.T. p. 40f.].

[7] 'Die Grundlagen der Mathematik' (with remarks by H. Weyl and an appendix by P. Bernays). *Abh. a. d. Math. Sem. d. Hamb. Univ.* **6** (1928), 65–92, p. 79.

[8] l.c., p. 66.

[9] 'Die logischen Grundlagen der Mathematik,' *Math. Ann.* **88** (1923), 151–165, p. 153.

[10] 'Über das Unendliche', *ibid.*, p. 179. For the 'method of ideals' discussed by Hilbert in connection with his theory of proof, see below p. 51ff.

[11] This interpretation, as we shall see, does indeed have consequences for mathematical theory itself, insofar as it leads to an attempt to prop up Cantor's theory of the non-denumerable infinite.

[12] Against this, Hilbert's collaborator P. Bernays ('Über Hilberts Gedanken zur Grundlegung der Arithmetik', *Jahresb. d. deuteschen Mathematikervereinigung* **31** (1922), 10–19, p. 16): "Where concepts are wanting a sign appears at the appropriate time. This is the methodological principle of Hilbert's theory." In considering this dictum, we see how strongly the symbolism of their own science fascinates those who handle it with the greatest skill, and how rightly we may be mistrustful of epistemological interpretations of their own specialized scientific work by even the most outstanding enquirers. By analogy we may point to the positivist view of scientific results by natural scientists themselves, who have led Husserl to say: "If natural science actually speaks, we gladly listen as disciples, but it is not always science that speaks when scientists speak; and certainly not, when they talk about 'natural philosophy' or the 'epistemology of natural science'." [*Ideen*, p. 38, E.T. p. 86]. Cf. also Aloys Müller, 'Über Zahlen als Zeichen', *Math. Ann.* **90** (1923), 153–158, and P. Bernays' reply to it, *ibid.*, pp. 159–163; also O. Becker, 'Mathematische Existenz', l.c., p. 453ff.

On Hilbert's symbolism and theory of proof, the following additional writings may be stressed:

D. Hilbert, 'Neubegründung der Mathematik'. First communication, *Abh. a. d. Math. Seminar. d. Hamb. Univ.* **I** (1922), 157–177.

D. Hilbert and W. Ackermann, *Grundlagen der theoretischen Logik*, Berlin 1928.

P. Bernays, 'Axiomatische Untersuchungen des Aussagenkalküls der "Principia Mathematica" ', *Math. Zeitschr.* **25** (1926), 305–320.

W. Ackermann, 'Begründung des "tertium non datur" mittels der Hilbertschen Theorie der Widerspruchsfreiheit', *Math. Ann.* **93** (1924), 1–36.

W. Ackermann, 'Zum Hilbertschen Aufbau der reellen Zahlen', *ibid.* **99** (1928), 118–133.

J. v. Neumann, 'Zur Hilbertschen Beweistheorie', *Math. Zeitschr.* **26** (1927), 1–46.

W. Dubislav, 'Elementarer Nachweis der Widerspruchslosigkeit des Logikkalküls' is to appear shortly in *Journal. f. Math.* **101**.

The ideas that J. König puts forward in his *Neue Grundlagen der Logik, Arithmetik und Mengenlehre*, Leipzig 1914, are in many points related to those of Hilbert on the foundations of mathematics.

[13] Cf. below p. 94ff.

[14] Cf. also the account of monomorphism of axiom systems, p. 60ff. below.

[15] On the history of the logical calculus, cf. G. Stammler, *Begriff, Urteil, Schluss*, Halle a. d. S. 1928, p. 83ff.

[16] Cf. Wittgenstein, l.c., 4.121: "Propositions show the logical form of reality. They display it." 4.1211: "Thus one proposition '*fa*' shows that the object *a* occurs in its sense, two propositions '*fa*' and '*ga*' show that the same object is mentioned in both of them."

[17] Cf. O. Neurath, 'Eindeutigkeit und Kommutativität des logischen Produktes "*ab*"', *Archiv. f. syst. Phil.* **15** (1909), 104–106, and 'Definitionsgleichheit und symbolische Gleichheit', *ibid.* **16** (1910), 142–144.

[18] *Der logische Aufbau der Welt*, p. 13 [E.T. p. 21], Cf. also his recent *Abriss der Logistik* (Schriften zur wissenschaftlichen Weltauffassung, vol. 2, Vienna 1929).

[19] According to a terminology that has been gaining ground more recently, many-one relations include one-one relations. This convention is used for example in *Principia Mathematica*.

[20] Cf. above p. 22.

[21] A further danger is that misconceptions that enter into the symbolism are thereby conserved and become quasi-legitimate. Wittgenstein's criticism of the symbolism in *Principia Mathematica*, the most thorough symbolic language we possess to date, should here be seen as a reminder to critical caution.

[22] 'Über das Unendliche', l.c., p. 174.

[23] 'Grundlagen der Mathematik', l.c., p. 79. In the same work he reduces existence theorems to a single axiom. "The source of pure existence theorems is the logical ϵ-axiom, on which in turn rests the construction of all ideal statements." (*ibid.*)

"The logical ϵ-axiom is $A(a) \rightarrow A(\epsilon A)$. Here $\epsilon(A)$ denotes a thing for which the statement $A(a)$ certainly holds if it holds for anything; we call ϵ the logical ϵ-function. To explain its role we observe the following :

"The ϵ-function is applied in threefold manner in formalist theory.

"i. By means of ϵ we can define 'all' and 'there is' . . .

"ii. If a statement $\mathfrak{A}$ holds for one and only one thing, then $\epsilon(\mathfrak{A})$ is the thing for which $\mathfrak{A}(a)$ holds. The ϵ-function thus enables us to resolve a statement $\mathfrak{A}(a)$ that holds for only one thing in the form $a = \epsilon(\mathfrak{A})$.

"iii. Besides, ϵ plays the role of a choice function, that is ,where $\mathfrak{A}(a)$ can apply to several things, $\epsilon(\mathfrak{A})$ is any one of these." (l.c., p. 67f.)

[21] Cf. especially Section I, note 17 concerning existential judgments as parts of judgments.

[25] These are in question in Hilbert's famous first ('theological') proof of the finitude of the complete system of invariants ('Über die Theorie der algebraischen Formen', *Math. Ann.* **36** (1890), 473–534, p. 521ff.).

[26] In his Vienna lecture 'Mathematik, Wissenschaft und Sprache,' printed in *Monatshefte für Mathematik und Physik* **36** (1929), 153–164.

[27] Here and in similar cases we speak of a 'postulate', understanding by it a demand made on thinking which it must satisfy in order to be 'correct thinking'. In fact, however, Brouwer's thesis means that 'non-constructible mathematical objects' are nonsense, and cannot be thought at all. The illusion of intentional reference to mathematical objects here arises only through attendant conceptions (cf. Husserl, *Logische Untersuchungen*, vol. II, 1. p. 61ff., E.T. p. 291ff.), which vanish when we try to make their 'proper' (central) sense precise.

This transposition of logical insights into demands on thinking is not without danger, since it easily leads to a merging of logical with psychological and anthropological aspects (cf. Husserl, *ibid.*, vol. I, p. 9ff., E.T. p. 58ff.).

[28] The main representatives of the older 'intuitionism' are Kronecker and Poincaré, and also Borel, Baire and Lebesgue. The name of this doctrine is intended to express the fundamental importance for mathematics that it attributes to the intuitive grasp of the series of natural numbers. Poincaré calls himself a 'pragmatist'.

[29] Amongst his writings cf. especially: *Das Kontinuum, Kritische Untersuchungen über die Grundlagen der Analysis*, Leipzig 1918; "Der circulus vitiosus in der heutigen Begründung der Analysis", *Jahresbericht d. Deutch. Math.-Ver.* **28** (1919), 85–92; 'Über die neue Grundlagenkrise der Mathematik', *Math. Zeitschr.* **10** (1921), 39–79; 'Randbemerkungen zu Hauptproblemen der Mathematik,' *ibid.* **20** (1924), 131–150; 'Die heutige Erkenntnislage in der Mathematik,' *Symposium* **1** (1925), 1–32, (also available as fascicule 3 of *Sonderdrucke des Symposion*); *Philosophie der Mathematik und Naturwissenschaft*, Munich and Berlin, 1927.

[30] Cf. works mentioned, and 'Das symbolische in der Mathematik', *Blätter fur deutsche Philosophie* **1** (1928), pp. 329–348.

[31] 'Intuitionistische Mengenlehre', *Jahresb. d. Deutsch. Math. Ver.* **28** (1919), 203–208 (also in *Kon. Akad. v. Wetensch. te Amsterdam, Proceedings* **23** (1920,) 949–954.) Cf. also the clear and intelligible formulation that H. Weyl has given to this idea in 'Über die neue Grundlagenkrise der Mathematik', l.c., p. 42: "The sense of a clear and unambiguously stipulated concept of an object may well serve in every case to assign to objects of the kind stated in the concept their sphere of existence; but it can by no means establish that the concept has a definite denotation, that it makes sense to speak of the existent objects falling under it as of some aggregate definite in itself, bounded, and ideally closed."

[32] 'Intuitionisme en formalisme', Groningen 1912, English translation in *Bull. of the Amer. Math. Soc.* **20** (1914), 81–96, p. 86.

[33] Cf. O. Becker, 'Mathematische Existenz', l.c., p. 448ff.

[34] Cf. 'Intuitionistische Zerlegung mathematischer Grundbegriffe', *Jahresber, d. Deutsche Math.-Ver.* **33** (1925), 251–256.

[35] However, on Brouwer's theory absurdity of absurdity of absurdity is equivalent to absurdity; likewise fourth level absurdity with second level.

[36] Objections to this terminology were raised especially by Weyl, 'Grundlagenkrise', l.c., p. 52; and by W. Burkamp, *Begriff und Beziehung*, Studien zur Grundlegung der Logik, Leipzig 1927, p. 129.

[37] Before we tackle this analysis, the following remark is important: It would be wrong to think that in formulating this question the decimal expansion of π had to be presupposed as a completed infinite totality. Rather, this statement, like meaningful

propositions about sequences as such, (cf. below p. 84), relates to a general term. Let us demonstrate this for the example of $\sqrt{2}$, which is in principle the same though technically simpler to formulate; that is, let us give a finite formulation of the proposition 'the decimal expansion of $\sqrt{2}$ contains a sequence 0, 1, 2 . . . 9'. We proceed as follows: let ν be an arbitrary natural number, and let $\nu' = a.10^r + b.10^{r-1} + \ldots + f.10^{r-s+9} + g.10^{r-s+8} + \ldots + p.10^{r-s} + \ldots + 10u + v$ (where $a, b, \ldots u, v, r, s,$ are natural numbers) be the biggest natural number whose square is less than 2.10^ν; then-the proposition '$f = 0$ and $g = 1$ and $h = 2 \ldots$ and $q = 9$' is consistent.

[38] Becker, too, recognized this clearly ('Mathematische Existenz', p. 508 note 1); but his attempt to save Brouwer's basic view in spite of this by means of phenomenological consideration seems to me unsuccessful.

[39] Taking the word in the widest sense.

[40] Cf. above p. 16, and below p. 151ff.

[41] Cf. above p. 46.

[42] Cf. for example 'Intuitionism and Formalism', l.c., p. 85: "However weak the position of intuitionism seemed to be after this period of mathematical development [discovery of non-Euclidean geometries], it has recovered by abandoning Kant's a-priority of space but adhering the more resolutely to the a-priority of time. This neo-intuitionism considers the falling apart of moments of life into qualitatively different parts, to be reunited only while remaining separated by time as the fundamental phenomenon of the human intellect, passing by abstracting from its emotional content into the fundamental phenomenon of mathematical thinking, the intuition of the bare two-oneness."

[43] Brouwer, with admirable drive and consistency, has furthermore built up a set theory and theory of functions without the principle of excluded middle, in which there is no room for the non-denumerable infinite of classical set theory. (For Brouwer's 'continuum' is quite different from Cantor's $2^{\aleph_0}$.) Cf. 'Zur Begründung der intuitionistischen Mathematik' I–III, *Math. Ann.* **93** (1925), 244–257; **95** (1926), 453–472; **96** (1927), 451–488. Also: 'Begründung der Funktionenlehre unabhängig vom logischen Satz vom "ausgeschlossenen Dritten" ', I, Amsterdam 1923. Concerning the connections between Brouwer's and the non-intuitionist set theory cf. Menger, 'Bemerkungen zu Grundlagenfragen, I. Über Verzweigungsmengen' and 'III. Über Potenzmengen', *Jahresber. d. Deutsch. Math.-Ver.* **37** (1928), 213–226 and 303–308.

[44] 'Philosophie der Mathematik und Naturwissenschaft', l.c., p. 16.

[45] It remained, however, doubtful whether all empty axiom systems are inconsistent. Meanwhile R. Carnap, in an as yet unpublished paper that he has let me see in manuscript, has shown that every provably empty axiom system is provably inconsistent.

[46] As M. Geiger has pointed out (*Systematische Axiomatik der Euklidischen Geometrie*, Augsburg 1924, p. 23ff.), it is better to speak of axioms being mutually 'underivable' rather than independent. For with an axiom's being underivable from others it is "quite compatible, that this axiom presupposes the existence of other axioms, if it is to figure as a meaningful proposition at all. For propositions about triangles to be meaningful there must for example be axioms from which it follows that straight lines can intersect in the first place; moreover it must be axiomatically guaranteed that they can have three intersections. Failing such axioms, there can be no triangles and therefore no axioms about them and so on. The axiom of congruence is thus not independent from the axioms stated, although it cannot be derived from them" (l.c., p. 27f.). This distinction is

important for a systematic axiomatics as postulated by Geiger, whose basic ideas we shall present in brief outline below. The requirement of independence can be made more stringent in various ways; as for example by means of the concept of 'complete independence' in E. H. Moore, *Introduction to a Form of General Analysis*, New Haven 1910, and E. V. Huntington, 'A New Set of Postulates for Betweenness with Proof of Complete Independence', *Transact. of the Amer. Math. Soc.* **26** (1924), 257–282; another example is given by R. Carnap in a forthcoming paper.

[47] Thus Hilbert and his disciples use the basic logical concepts 'or' and 'not', although Sheffer has shown that they can both be replaced by the relation of incompatibility.

[48] Cf. A Fraenkel, *Einleitung in die Mengenlehre*, 3rd edition, Berlin 1928, p. 347ff.

[49] Cf. the following remarks by E. Husserl about 'definite multiplicity' ('mathematical multiplicity in the pregnant sense').

"This multiplicity is characterized by the fact that a finite number of concepts and propositions (possibly to be drawn from the nature of the field in question) completely and unambiguously determines the totality of all possible formations of the field in the way of purely analytic necessity, so that in principle nothing remains open.

"Alternatively, we may say that such a multiplicity has the special property of allowing 'exhaustive mathematical definition'. The 'definition' lies in the system of axiomatic concepts and axioms, and the 'mathematical-exhaustive' aspect in that the defining assertions imply the biggest conceivable restriction for the multiplicity – nothing remains undetermined.

"An equivalent to the concept of a definite multiplicity resides in the following sentences: every proposition to be formed in whatever logical form from the axiomatic concepts indicated is either a purely formal logical consequence of the axioms, or a non-consequence, that is, inconsistent with all the axioms, so that its contradictory would be a formal logical consequence of the axioms. In a mathematically definite multiplicity the concepts 'true' and 'formal logical consequence of the axioms' are equivalent, and likewise the concepts 'false' and 'formal logical non-consequence of the axioms' ". (*Ideen*, p. 135f., E.T. p. 204f.) Cf. also the analysis in *Formale und Transzendentale Logik*, p. 78ff.

[50] Cf. also below p. 153ff.

[51] l.c., p. 12.

[52] p. 34f.

[53] An improper definition consists in "that an object is defined by a relation in which it stands to other objects that are supposed to be already known" (l.c., p. 15).

[54] Geiger distinguishes between axioms that posit existence and axioms that characterize. Amongst the former are those that give the number of element systems and relation systems. "In contrast . . . we have characterizing axioms that give closer determination to the character of relations and through these to the character of the elements. An axiom of this kind would be for example the following: two different straight lines intersect only in one point. Such an axiom does not indicate how many relations there are between points and straight lines, but further characterizes the relation of intersection that exists between them by indicating between how many of them it can exist" (l.c., p. 32).

[55] The domain of things that forms the special object of investigation is called 'qualified'.

III

NATURAL NUMBER AND SET

If, following the path indicated by our considerations so far, we proceed to define the concept of natural number, we must begin with the description of the state of affairs in which numbers are first given, the 'model' of numbers; next, we must isolate numbers by abstracting them from that state of affairs. This latter is the process of counting, about which we can make two preliminary remarks: (1) any arbitrary objects may be counted, thus insights about the number concept gained by descriptively analysing the counting process hold independently of what we happen to be counting; (2) no new property accrues to objects through being counted. The second point needs some elucidation.

What there being twelve apostles means, as Husserl remarks in his *Logical Investigations*, is not that each apostle is 'twelve'; if twelve paintings hang on the walls of a room, none is affected in its own character by there being eleven others. Each of them can be viewed separately in succession without their being counted, that is, without their being given an ordinal index relative to those previously viewed which would fix, with regard to those viewed earlier or later, their temporal position in the viewer's consciousness.

If we do fix such an order, then one painting will be the first, second, third, . . . twelfth; but which is which depends on the viewing sequence, which is by no means given in advance, say by nearness in space. Moreover differences in the manner of viewing may arise by 'grouping together' certain pictures looked at in rapid succession, while greater time intervals separate them from the viewing of the others.

What, then, is the fixed point or invariant of all these variations? We recognize that it lies in the ordinal position of the 'last' element. In our example this is always the 'twelfth', however the viewings are grouped. In order to state this mathematically vital state of affairs, let us introduce a precise formulation.

Given various distinct things (T) and certain other things different from the former and from each other, which we shall call signs (S), the following stipulations are to hold: to each T we assign one and only one S, as far as the supply of the latter lasts. Which S is assigned first is fixed, and so is which S will be assigned after any assigned S; but we leave unfixed which S is assigned to which T (the manner of the one-one assignment is arbitrary). With regard to the various modes of such assignment, what remains invariant is the S assigned to the last T (or, if we are short of S, the S assigned to the last T encompassed by the assignment). If we assume that there are enough S and that the T to be denoted are unambiguously fixed, this determines one and only one S, which is the sign of whatever T happens to be the last.[1]

If in a certain such process with a given order of signs the last sign used is n, we can describe this state of affairs by saying that 'the sign n corresponds to the "totality" of counted things'.

However, we must not interpret this terminology as if by fixing the things to be encompassed by such a denoting process we had constituted, independently of that process, a 'totality of things' whose 'properties qua totality' were logically prior to the result of the denoting process (that is, of ascertaining the sign of the last thing denoted); all this on the grounds that the denoting process just described contained an order of the things counted, while the concept of totality as such did not.

Indeed, the concept of totality cannot be meaningfully described except if correlated with such an ordering process, which encompasses the various things and thus 'collects' them. Since, as already anticipated, the denoting process just analysed is simply the process of counting, our stipulations can be translated into mathematical language as follows: the statement that a set of n things is being counted simply states that, whatever the arrangement of the things counted, the last will be the nth.

Two important consequences follow: (1) since the cardinal number is simply the ordinal number (positional sign) of the last element of the things counted under arbitrary arrangement, this resolves the controversy fought by generations of mathematicians as regards the logical priority of cardinal or ordinal; (2) in defining the concept of number we do not require the concept of a set.[2]

In order not to accumulate difficulties, let us postpone further analyses connected with the concept of a set, resuming them only after we have accurately fixed the concept of natural number. We can, however, at once determine the common origin of these two mental duplications: it lies in the mistaken treatment as independent of invariances that in fact are meaningful only with respect to a certain domain of variation, a situation analysed in detail in Section I. In the present case, that domain is formed by the different possible ways of arranging n objects: with regard to them the ordinal index of the 'last' object remains invariant, if the order of denotation is fixed; it is always the *n*th. This invariance, let us emphasize again, means only logical independence of a particular kind of arrangement, but not separability from order as such.

If nevertheless we assume that there is a number of objects independently of the ordinal number, further speculation leads directly to a 'carrier' of that number (of objects), namely the set. Of course this is only one of the two main reflective origins of the concept of a set, while the other (likewise described in Section I) must be sought in the fusion of individual and specific universality, about which more later.

Proceeding now in our analysis of the number concept, let us consider another question that is closely allied to the relation between ordinal and cardinal numbers, namely the relation of one-one correspondence between the elements of two sets and the order of the elements within each set.

Russell, following G. Frege, asserts that the former process is independent of the latter, citing the illustrative example of the number of men and women monogamously married.[3] He says on the basis of the attendant one-one correspondence alone we can obviously ascertain that the number of married men is equal to the number of married women, without first having to establish some sequence of them.

However, this argument is not valid, for the concept of 'being equal in number' (or, as Russell says, of 'similarity') without reference to any order at all does not yield the sense generally connected with it and intended by Russell; only the kind of order is arbitrary. Therefore the state of affairs that Russell is aiming at must be formulated correctly as follows.

Let there be distinct objects A, B, C, . . . and a, b, c, . . . with a one-one correlation fixed between corresponding members of the two sets, all objects of both being encompassed. Whatever the arrangement of either set, both will yield the same number, that is, ordinal index of the last element.

These Russellian theses, as indeed the doctrine of the primacy of cardinals, of which they are the most rigorous formulation, have been most strongly opposed by intuitionists, most recently by H. Weyl:[4] ". . . the possibility of pairing, which is mentioned in the criterion of equality of number, can be tested only if the acts of correlation are performed successively in an ordered temporal sequence so that the elements of both sets will thereby be themselves ordered. If the comparison of two sets is abstractly severed into determinations of the number of each set and subsequent comparison of the numbers, it is thus indispensable to order the individual sets themselves, by exhibiting one element after another in time . . ."

However, these comments, obvious though they might seem at first blush, conceal a new difficulty of principle, arising from the part played in them by time.[5]

The problem of the connection between time and number, which Kant,[6] above all other great philosophers of the past, laboured to resolve, essentially springs from the following dilemma of thought: on the one hand, counting, which is doubtless closely linked with number, is a process in time; but on the other hand the concept of time evidently does not enter number theory or the theorems of arithmetic: looking at any such theorem we find no trace of a relation of time.

If, however, we eliminate the time element from the definition of number, we are easily tempted to treat it as a kind of occult quality of 'totalities of things',[7] as a 'quality of the set' of items; or, if in place of quality we put a class of things (so avoiding difficulties of abstraction), to treat number as a class of totalities of things. If next, by fusing individual with specific universality, a totality of individual objects in the same way as a property of things, is denoted as a 'class' or 'set', we obtain the definition of natural number given by Russell, following Frege,[8] namely: the number of a class is the class of all classes similar to it. For example, the number two must be defined as the class of all pairs.[9] The difficulties that Russell becomes involved in when trying consistently to implement this conception show up especially in his theory of irrational (real) numbers.

From these observations it then follows that in trying to find an exact definition of natural numbers (that is, of the cognitive object with which mathematicians actually deal under that name), our method is largely prescribed, in that we must use neither the concept of time nor that of a totality of things or of a property of a complex (set, class and so on). However, the general insights gained in Section I concerning the way abstraction leads to the formal sphere will lead to a solution of the problem. Numbers turn out to be logical abstractions of the counting process conceived as capable of being continued without bounds. This, we note, adds an aspect of 'idealization' to the aspect of abstraction in 'deriving' number from the counting process, since every actual such process has an upper bound. This is important for determining the relation between logic and mathematics (to be undertaken at the end of this section).

In order to grasp the above thesis, let us make clear what it means, within the framework of the counting process, that an object is, say, 'the third'. This simply means that a 'second' object has preceded it; the 'second' is that which has been preceded by a 'first', while the 'first' is determined by lacking a precursor.

The structure of this relation, its logical abstraction, is obtained, in the sense of the observations of Section I, isolating the incompatibility relations contained in it, that is, by 'putting into brackets' the phenomenal aspect that consists in temporal sequence.

Thus we obtain: something is a 'third' if it is incompatible with there not being a 'second', which 'second' in turn presupposes a 'first'; but this does not yet sufficiently characterize the 'third' thing: for a fourth, fifth, sixth and so on also presuppose a second and a first. However, what characterizes the third is just that nothing is presupposed except a second and a first. Thus every number is a logical singularity, unambiguously defined by certain incompatibility relations, which it shares with all numbers greater than it but which for it alone are linked with the exclusion of any other incompatibility relations. Thus the number series presents itself as a boundless superposition of incompatibility relations on a 'first' element, the immediate successor of a number including just one such relation in addition to those that define its 'immediate precursor'.

Every natural number is logically unambiguously fixed by incompatibility relations, which we shall denote by the term 'logical singularity'. Any further stipulation of incompatibility relations would thus be either redundant or contradictory.

One restriction is needed: the remark just made holds only insofar as we exclude any relation between the terms in question that cuts across the incompatibility relation: otherwise we could stipulate new incompatibility relations consistent with the old ones, by reference to what does or does not belong to the domain of relations that cut across.[10]

In defining natural numbers we must therefore exclude such relations or the terms defined by them. The following definition results:[11] We denote as natural numbers the elements of the structure determined by the following stipulations and by these alone:[12]

(1) There is one and only one element with whose presence the absence of no other element is incompatible.

(2) For every element N_m there is one and only one element N_n with whose presence the absence of N_m is incompatible, while the presence of N_n is in addition incompatible with the absence of only such N_m different from N_n whose absence is also incompatible with the presence of N_m.

(3) The relation between N_m and N_n determined by (2) is incompatible with any other element being related in like manner with N_n.

This 'presence' is not to be understood as if the numbers as such were 'there'. That a particular natural number is present is merely to mean that it is thought of as assigned to an object according to the above formal rules.

Alternatively, we can describe this state of affairs by starting from the number signs, in which case the 'presence' of a number means the use of a particular number sign within the framework of the unambiguous stipulated rules of use. It would be a gross misinterpretation if, inspired by an overwrought realism, one were to take 'presence' to imply that 'arithmetical existence' of large numbers or meaningful operations with them depended on whether there were sufficiently many things 'in the world'. A rejection of this view follows unambiguously from our observation that the concept of the number series contains an idealization.

How vital it is, for the theory of mathematical method, to be clear about the connection between logico-mathematical abstractions and their 'model', reveals itself likewise in regard to the controversy of method between formalist and intuitionist doctrine. Indeed one of the four basic theses in which Brouwer has recently summarized the essence of his critique of method, states "that the justification (as to content) of formalist mathematics by the proof of its consistency contains a vicious circle, because this justification rests on the correctness (as to content) of the statement that from the consistency of a proposition its correctness follows, that is, on the correctness of the principle of excluded middle".[13]

Now we entirely agree with Brouwer that the question concerning the correctness as to content, that is, the cognitive character of mathematics, can be answered only by an 'intuition', that is, an insight about the world. However, the formalists will reply that within the framework of their doctrine they have never answered nor even asked such a question, since this question is an epistemological question and therefore lies outside mathematics. Moreover, they can assert that it is precisely the epistemological analysis based on the 'original intuition' of the temporal model of the number series that exhibits contradiction as the only barrier to formal connections. So far so good; but we must not forget that the formal sphere is itself part of the 'world' and can therefore not be arbitrarily extended by the introduction of new symbols. The attempt at such an extension by means of the thesis of mathematical existence of the non-denumerable infinite was indeed the main reason for Brouwer's critique, as we have seen.

There is a further important consequence for foundational problems in mathematics that follows from a grasp of the relation between natural numbers and their temporal model. For once we recognize that natural numbers are themselves formal concepts, it must seem hopeless to build up, alongside the 'contentful mathematics' of natural numbers, a 'formal mathematics' in which to rescue the non-denumerable infinite of set theory. For the sense of this last 'formalisation', which even leaves aside the meaning of the 'logical constants' ('not', 'and' and so on), lies merely, as we have shown in Section I, in setting up formation rules that 'fit' the transformations in the sphere of logical constants.

Given these observations it is obvious that it is only by reference to the model of the number series that we can tell what is the epistemological significance of the series' being infinite. Note that this is the problem whether this assumption has the character of knowledge, that is whether and how it says something about the world. It then becomes plain that the model of this infinity appears as boundlessness in principle in time. A temporally ordered chain of events, of which counting is an example, is indeed incomplete in the sense that we can always form a new chain that contains all events of the previous chain and one further event as well.

This is not to say that an infinite series of events could be 'present'; rather it means that every series of events that is present is finite (characterized by a particular natural number), but all events can never be present. The 'model' of infinity is therefore a rule-governed process that goes ahead but cannot be completed, exactly in Kant's sense.[14] Adding a further element to the series is always the same process (subject to the same rule), however many elements the series already contained. The formal abstraction of this process is the superposition of incompatibility relations which has a beginning but no end.

Since, according to what we saw in Section I, no formal concept is alien to arithmetic and unbounded iterability of any formal concept is guaranteed by the infinity of the number series, arithmetic forms the universal schema for all conceivable formal systems. It is therefore impossible to construct a formal system of relations that could not be isomorphically represented on a proper or improper sub-system of arithmetic.

In order to go on to one of the main problems of the theory of natural numbers, namely that of analysing the principle of complete induction, let us examine the extent to which our axiomatic determination of natural numbers agrees with and differs from the classical axiom system of Peano.

Peano[15] uses three basic concepts, '0', 'number', 'successor', and stipulates the relations between them by means of the following axioms: 1. 0 is a number. 2. The successor of any number is a number. 3. No two numbers have the same successor. 4. 0 is not a successor. 5. Any property of 0 that belongs to the successor of any number that has it belongs to all numbers.

These five axioms are so fixed that with the addition of the principles of pure logic we can develop from them the whole theory of natural numbers. If we compare them with our axioms, we see that the concept of a first number (the 0 of Peano) and a successor acquire their precise logical meaning through our axioms in the first place, being unambiguously fixed by incompatibility relations. For the rest, our three axioms correspond to Peano's first four.

What, then, about his fifth axiom, which has been rather infelicitously called the 'principle of complete induction' or 'mathematical induction', or, more appropriately the 'law of inference from n to $n + 1$'?

This law states that any property of 0 (that is, of the first element of a sequence for which the first four Peano axioms hold) invariant to the relation of succession belongs to any element of the sequence whatever. If we ask under what conditions this law applies, we recognize as the necessary and sufficient condition that starting from the first term any term 'can be reached in finitely many steps'. In that case we can, for any arbitrary term, verify the law by a chain of inferences as follows:

Major premiss: if an element has the property P, so has its successor. Minor premiss: the nth element has the property P. Conclusion: the $(n + 1)$th element has the property P.[17]

However, the concept of 'can be reached in finitely many steps' is evidently not a purely logical one, since it contains a time aspect. The problem of complete induction thus lies in laying bare the logical abstraction of 'can be finitely reached'.

For the sake of perfect clarity, let us consider a series which satisfies the first four Peano axioms but not the fifth, namely

$$-1, -\tfrac{1}{2}, -\tfrac{1}{4}, -\tfrac{1}{8} \ldots \tfrac{1}{8}, \tfrac{1}{4}, \tfrac{1}{2}, 1, 2, 4, \ldots$$ [18]

This series satisfies the first four Peano axioms because it has a first term and every term and its successor stand in mutual one-one correlation, but it does not satisfy the principle of complete induction because the first term has 'properties' that, though invariant to the relation of succession, do not belong to all terms of the series, for example the 'property' of being a negative number. This is because the series is not formed by the relation of succession alone, but by its combination with other distinguishing features, such as being positive and negative numbers.

That the positive terms 'cannot be reached' from the first term simply means that they cannot be defined by means of it and the relation of succession. The time aspect drops out completely here, and the 'infinite number' of steps that on imprecise formulation lie between the negative and positive terms of the series is merely an incorrect description of the fact that within the series there are divisions that cannot be formulated by means of the relation of succession. To exclude stipulations leading to such divisions, as we have done in our definition of natural numbers, is thus necessary and sufficient if elements of a structure subject to the first four Peano axioms are to obey the fifth one as well.

Further, to see through the entire position, we must clarify what are the elements on which the main difficulties that stood in the way of a solution to the problem rest. This is important for two reasons: first, in its own right, because the principle of complete induction is of fundamental importance to the theory of mathematics; and secondly, for the following investigations, because we have here a paradigm of a transfinite problem; we may therefore expect that by critically analysing the misconceptions that prevail in this field we can discover the sources of misconceptions besetting the treatment of other transfinite problems.

Two circumstances pre-eminently barred the way to a solution of our problem. First, because of the fusion of individual with specific universality, numbers were regarded as an infinite totality of individual objects which, as it were, gained connection only afterwards, through regularities holding between them. Secondly, mathematicians thought that in 'transfinite numbers' they had found numbers that could not be reached from the first number in finitely many steps.

The second thesis, to be treated in Section V when we shall analyse the theory of well-ordering, did not become important for an understanding of the problem until a later stage, after Cantor had created his set theory. The consequences of the first error may, however, be clarified at once.

If one holds that statements about 'all numbers' directly relate to an infinite totality of 'individuals' (singularities), it follows that one will see in the law of complete induction the key of a specific procedure superior to syllogism, because the latter had to remain confined to a finite number of inferences, in contrast with mathematical induction. The foremost supporter of this view amongst modern mathematicians is H. Poincaré. He takes the inference from n to $n + 1$ as an axiom that constitutes the methodological peculiarity of mathematics and asserts that any attempt to prove this law by pure logic must lead to a circular argument, since in any such attempt tacit use is made of the law to be proved.[19] This view of the great French mathematician contains a grain of truth insofar as this law is indeed a specific feature of natural numbers. However, it is to misdescribe the state of affairs that exists here if we see in the law an extension of logical method, setting it over against the analytic judgments of logic as an *a priori* synthetic judgment.

The 'synthetic aspect', the 'intuition' or 'original intuition', lies elsewhere, as we have already noticed, namely in the model of the superposition of incompatibility relations, that is, in the process of counting. In the 'intuition' of the counting process we grasp that the progression of a counting act to the next number is independent of what has been counted before, and the logical abstraction (the structure) of this boundless progression of the counting process is the 'infinite series' of natural numbers. This series therefore is neither something that is successively becoming, since we have abstracted from the time aspect, nor yet a totality of infinitely many singularities existing as such; but it is the 'field' (domain) of the boundless superposition of incompatibilities. That the principle of complete induction holds is logically fixed by the constitutive principle of the series of natural numbers. It is the validity of the conclusion of an inference with arbitrarily many terms and in no way contains a peculiar procedure of its own.

If we now ask what is the cognitive value of the above analyses leading to the result that our definition of natural numbers is not only equivalent to Peano's axiomatic system but agrees with it as regards procedure,[20] it is evident that the advantages of our definition are not to be sought in 'technical' aspects. Rather, they consist in the fact that these analyses make completely precise what is the meaning of Peano's axioms, both as regards the basic concepts 0, number, successor, and, independently of these, the concept of the principle of complete induction. For since our definition of natural numbers is obtained in a perspicuous manner by an analysis of the structure of the counting process, we appear to have answered the question as to the 'nature' of number, that is, its place in the 'edifice of the world' ('system of knowledge'). It would be completely to misunderstand the nature of this problem if one were to interpret such a question as an appeal to some vague intuition and therefore to reject it. On the contrary, this question merely imposes the task of making precise what is the theme of number theory, that is, of becoming fully aware of what one is talking about in doing arithmetic or just calculating. Once this is grasped, the fact that mathematics is in principle fit to serve the purposes of natural science loses its mystery: for from seeing how the number series is connected with the order of succession of the counting process, the mode of connection of numbers with the world issues as a result.[21]

Moreover, this sheds light on the foundation problems of mathematics itself. First of all, it is easy to establish that our definition of natural numbers is consistent; for since it is simply a description of the formal skeleton of the counting process conceived as boundless, that is, its abstraction from the phenomenal aspect of time sequence, any contradiction arising in it would imply that an unambiguous counting process was impossible.

Likewise our analysis of the counting process immediately yields the 'completeness' of our definition in the sense that it contains everything that can be formally stated about the counting process. From this it follows further that our definition is 'complete' in the sense of Section II.

This can be shown as follows: that the formal system of natural numbers set up by our definition is non-branchable and monomorphic is at all events established beyond doubt, if we can show that it covers the whole formal domain, since in that case there is no room left for branching or indeed any further specifications; but this is the case here.

For from what we saw in Sections I and II it follows first that the formal domain must be regarded as completely described, that is, different from all other domains, since generalizations merely mean partial indeterminations (variabilities). Moreover, we saw in Section I that the formal domain can be described (after distinguishing between variables and logical constants) with the help of the one logical constant 'incompatible with'. With respect to this concept the natural numbers are defined in such a way that in principle nothing remains open. For the definition of each individual natural number consists in the setting up of certain incompatibility relations all other such relations being excluded, so that each such schema determines exactly one natural number as the 'bearer' of these relations.

Accordingly, the natural numbers are the singularities of the formal domain and their adequate description must imply all that can be said within the formal domain. From this follows the uniqueness of arithmetic as against the multiplicity of the systems of relations called 'geometries', whose universal schema (as indeed the universal schema of any possible system of relations) is represented by arithmetic, which shows itself in the fact that any arbitrary geometry can be represented on systems of numbers.

In Section VI we shall show that the assertions that arithmetic is non-branchable and decidable coincide.

Our result shows a certain affinity with the view of Russell, who defines the natural numbers with the help of the principle of complete induction.[22] He comments as follows:[23] "... mathematical induction is a definition, not a principle. There are some numbers to which it can be applied, and there are others ... to which it cannot be applied. We define the 'natural numbers' as those to which proofs by mathematical induction can be applied, i.e. as those that possess all inductive properties. It follows that such proofs can be applied to the natural numbers not in virtue of any mysterious intuition or axiom or principle, but as a purely verbal proposition. If 'quadrupeds' are defined as animals having four legs, it will follow that animals that have four legs are quadrupeds; and the case of numbers that obey mathematical induction is exactly similar".

We agree with Russell that the principle of complete induction contains no extra-logical procedure and that it is unambiguously fixed with the law of formation (the definition) of the natural numbers.

As against this, we diverge on the following points: (1) In order to define the concept of natural number Russell uses the concept of a set (class), which, as we shall show, must be rejected as ambiguous. (2) We do not agree with his view of non-inductive series, which takes its bearings from the doctrine of well-ordering in set theory. These two points of difference will emerge more clearly still in the following analyses.

Although the preceding investigations show that the principle of complete induction is not an *a priori* synthetic principle but a logical consequence of the law of formation of natural numbers, this law itself, which sets up a boundless superposition of incompatibility relations, seems to open up a way towards infinite multiplicities. The problem of alleged operations in the infinite domain will be our main concern in Section IV; but as an indispensable pre-requisite for understanding this we must in principle take stock of the character of multiplicities or sets, in order to be able to grasp the true meaning of mathematical statements that seem to relate to infinite sets. These are statements of the form 'all numbers (or all numbers having a particular property P_1) have the property P_2' and 'there are (amongst all numbers) numbers having the property P_2'.

In analysing the counting process earlier, we saw that in order to define number we do not need the definition of a set. Now we must enquire whether the term 'set' can be given an unambiguous sense at all.

Let us begin with G. Cantor's definition (although it has by now been superseded): "By a 'set' we are to understand any collection into a whole

M of definite and separate objects *m* of our intuition or our thought. These objects are called the 'elements' of *M*."[24]

Our analysis starts with giving a precise account of the concepts 'definite' and 'separate'. The requirement of definiteness is satisfied if for every object it has been ascertained whether or not it is an element of the set in question. As regards separateness, this is fulfilled if for any two elements *A* and *B* it has been ascertained whether they are conceptually identical or different.

What is decisive is that on this definition a set is completely defined by its elements, so that sets are equal if and only if they contain the same elements.

The 'collection of objects into a whole', according to the view prevailing in set theory, can be performed in two ways: first, by counting objects and thus 'collecting' them through entering them into the counting process; and secondly, by collecting 'all objects having a certain property' on the basis of that property. This produces a dangerous ambiguity in the concept of a set.

The genesis of this ambiguity probably was that what is common to the objects to be counted, namely the entering of each of them into the counting process, was hypostatized into a unity of a higher kind, which affords greater convenience in the mode of expression, as in other instances of hypostatizing. This 'collection' of objects into an 'object of higher order' seemed moreover intuitively justified by the fact that the intuitive objects to be counted appear from the outset as 'collected' in a certain region of space.

If, however, we take the word 'set' in the second sense indicated and fail to keep 'objects of intuition' (individual empirical things) and 'objects of thought' (numbers) strictly apart, we obtain the analogy between properties of corporeal things and 'properties' of numbers.

In fact, however, the 'properties' of numbers turn out to be 'relative incompatibility relations'. Take for example the divisibility of integers. A number z is divisible by the number n if it is the biggest of n numbers each of which save the smallest exceeds the next smallest by the smallest. In mathematics we usually put it thus: 'a number z is divisible by n if there is a number m which multiplied by n gives z'. The sense of what we have shown in this example applies to any 'mathematical property'.

Consider next some examples in which we shall show how propositions that are usually formulated with the help of the concept of a set will be reformulated on elimination of this concept.

(1) To the proposition 'the set of all natural numbers contains the set of all prime numbers' corresponds the statement 'if something is a prime number, then it is a natural number.'

(2) To the proposition 'the set of all natural numbers a, b, c, $n > 2$ contains no Fermat quadruplet of numbers' corresponds the statement 'if a, b, c and n are natural numbers and $n > 2$, then they are connected by the relation $a^n + b^n - c^n \neq 0$.

(3) To the proposition 'the set of all natural numbers is infinite' corresponds the statement 'any arbitrary natural number n determines a number $n + 1$ different from the numbers required to determine it'.

(4) To the proposition 'the set of natural numbers between 5 and 12 is smaller than the set of natural numbers between 20 and 30' corresponds the statement 'between 5 and 12 there are fewer natural numbers than between 20 and 30'.

These examples are particularly simple, but more complicated ones would offer no new aspects of principle, nor would they impair the result which will be further confirmed in later sections, namely the following: in order to formulate legitimate mathematical statements we do not require the concept of a set, but where that concept (or the corresponding concept of a 'mathematical property') seems indispensable, there we are confronted with meaningless pseudo-mathematical propositions.

If the concept of a set can here be eliminated, it still might perhaps be required where it appears in iterated form. We shall now show that it can be eliminated in these cases too. Let us demonstrate this for the simplest example of iteration, namely for the concept of a set of sets. The argument can easily be transferred to further iterations.

Consider the following simple example: the set of sets [1, 2, 3, 6, 8], [4, 5, 6], [7, 9] contains three elements. What cognitive fact is expressed by this proposition? To understand this, let us first resolve the first of the two concepts of set, leaving the second one untransformed. That is, we ask ourselves: what is here meant by a '*set* of sets'?

On the basis of our previous findings this is merely another expression for the unambiguous assignment of natural numbers to the sets and vice versa. The second question is then what are these numbers actually assigned to, or, putting it differently, what aspect achieves the 'collection' of the various elements into sets.

This 'collecting aspect' consists in assigning the same number to certain elements in the correlation by which a 'set of sets' is determined. To the concept of a set of sets in our example there corresponds the following schema:

$$\left\{\begin{bmatrix}1, 1, 1, 1, 1\\ 1, 2, 3, 6, 8\end{bmatrix}, \quad \begin{bmatrix}2, 2, 2,\\ 4, 5, 6,\end{bmatrix} \quad \begin{bmatrix}3, 3\\ 7, 9\end{bmatrix}\right\}$$

The iteration of sets accordingly amounts to a superposition of such correlations.

If we further recall that the counting of objects consists in the unambiguous one-one assignment of ordinal indices and that the result of the counting process is independent of the kind of objects being counted, it is clear that ordinal indices themselves may be counted too. Only we must not forget that no visual or acoustic phenomenon is as such an ordinal index; it is an ordinal index only insofar as it has an ordering function within a thinking process. If now we count ordinal indices, then within the counting process they function not as ordinal indices but as 'arbitrary objects': nothing would be changed in the manner and result of this counting process, if instead of each individual ordinal index some other object was entered into that process.

This observation is important above all for preventing misinterpretation where the signs counted are the same in writing or sound as those by means of which we are counting; for that which is counted and number do not coincide here either. The decisive aspect, however, lies in the following observation: the 'collection of numbers into a set' as just described and illustrated in the above example, by assigning the same number to them, is evidently an empirical fact, and the same holds for the formation of a set of sets on the basis of such collections, and so on; this does not determine relations internal to mathematics.

If, however, we take the concept 'set' in its second sense as 'mathematical property', we cannot superimpose sets in such a way that there could be 'properties' of properties, 'properties' of 'properties' of properties and so on, as was shown in Section I. Yet we tend to overlook this because of the ambiguity of the concept of a set, so that we try to perform such incapsulations of 'sets' even where the superpositions of counting processes just described are not present, which leads to the formation of the extended functional calculus.

At this point the merging of two heterogeneous domains in the concept of a 'set' leads to mathematical absurdities, above all in the formation of the concept of 'set of all subsets of the set of all natural numbers'.

The formation of this concept needs closer inspection. The definition of a subset is as follows: a set M' is called a subset of a set M, if every element of M' is also an element of M. If M' is identical with M, then M' is called an improper subset of M. For example, the 'set of the numbers 1, 4, 17' and the 'set of prime numbers' are to be denoted as subsets of the set of natural numbers, which simply means that 1, 4 and 17, and every prime number (we are obviously here not considering prime ideals) is a natural number. By describing such mathematical facts with the help of the concept of a subset we are misled into setting up a property of 'being a subset of the set of natural numbers' and then assigning to this property a 'set of all subsets of the set of natural numbers', in the sense of the principle of comprehension, without regard to whether the individual subsets of the set of all natural numbers (which are their elements) are otherwise determined. On this concept of a 'set of all subsets of the set of natural numbers' Cantor based his doctrine of the non-denumerable infinite, of a sequence of orders of transfinite cardinal numbers and of transfinite classes of numbers. These we shall have to analyse more closely in Section V.

Against this misinterpretation we must emphatically point out that the concept of a set of all subsets of the set of natural numbers (as follows from the definition of a subset) is a subsidiary concept, presupposing other determinations on the basis of which the 'quality of being a subset' is attributed; therefore it is not permissible to operate with the concept of 'arbitrary subset of the set of natural numbers' or with the concept of 'set of all subsets of the set of natural numbers'. To this objection one might reply that, as shown previously in Section II, mathematical facts must be taken as existing independently of being discovered, so that we may after all operate with the concept of the 'existence' of the totality of such facts, even if the definition of each individual element of the totality is in principle excluded. This reply is however invalid, because we cannot speak of an actually existing infinite totality of natural numbers, let alone of a totality of 'properties' of natural numbers existing alongside them. This observation, which we made earlier, will receive further attention below.

The clearest and most profound insight into this state of affairs is probably due to Wittgenstein, but even Poincaré and Borel, and above all Weyl, show an attitude that is in principle correct.

Linked with the concept of a set as viewed in Cantor's set theory there are, as already mentioned in the introduction, a number of antinomies called the 'paradoxes of set theory' which mathematicians have found it very difficult to eliminate.

It is thus understandable that the enquirer who has fought most vigourously against these paradoxes and has perhaps struggled more than anyone with the problems of the foundations of mathematics, namely Bertrand Russell, has begun to have doubts as to the indispensable character of the concept of a set. Russell clearly expressed these doubts in his *Introduction to Mathematical Philosophy*, which was written during the 1914–18 war; on p. 184 he says of classes that they are probably no more than "symbolic fictions", and on p. 191 he observes that "the axiom of reducibility involves all that is really essential in the theory of classes". Since then his doubts seem to have been greatly strengthened, especially under the impact of the investigations of Wittgenstein and L. Chwistek,[25] for Russell now feels the defects of his reducibility axiom even more clearly than before. In the preface to the second edition of *Principia Mathematica*, p. XIV, he says: "One point in regard to which improvement is obviously desirable is the axiom of reducibility." "This axiom has a purely pragmatic justification: it leads to the desired results and to no others. But clearly it is not the sort of axiom with which we can rest content." Now that Russell's profound analyses in the theory of mathematics are fortunately becoming appreciated by speakers of German, it is doubly important to point with emphasis to the defects in Russell's theory that are caused by the ambiguity in the concept of a set.

In the removal of the confusions concerning the concept of a set those surrounding the concept of a sequence likewise disappear. A sequence is defined by a law by which a certain number is unambiguously assigned to every natural number. Such a law of formation, however, does not define a transfinite totality but merely an ordering relation between numbers having certain 'properties'. A statement about the 'general term of a sequence' therefore does not mean the mathematical transformation of a relation that originally held for an infinite multiplicity of individual elements, but it is the adequate expression for the general character of that relation, its application to a particular number being by contrast a logically later event. Nor does an understanding of the concepts 'sequence of sequences of numbers' or 'sequence of sequences of sequences of numbers', and so on, present any greater difficulties. By this we mean rules by which certain numbers are unambiguously assigned to ordered pairs of natural numbers or to ordered triplets of natural numbers, and so on. In the analysis of irrational numbers in Section IV and in that of transfinite ordinal numbers in Section V we shall recognize that propositions about infinite sets, sets of sets, and so on, must be transformed into propositions about sequences, sequences of sequences and so on, whereby their seemingly transfinite character vanishes. On the contrary it is not admissible to regard a sequence as a totality, as for example Cantor does in his definition of irrational numbers by means of fundamental series (cf. below p. 104).

If we are quite clear in principle on this and we moreover shun any temporal interpretations that might be psychologically suggested by the term 'sequence', it will be in the interest of practical simplicity of linguistic expression to use that term without qualms.

We have seen how dangerous it is to operate with the concept of a set if we are not entirely clear about the meaning of the symbols (or if their meaning is not precise, in Leibniz' sense); this danger stands out perhaps even more starkly in the so-called extensions of the number concept, which are of interest to us above all because they lead to the formation of the concept of irrational numbers and real numbers, which play an important part in the problems of the transfinite.

In carrying out 'extensions of the domain of numbers' one often starts from the operations inverse to the basic operations, as we shall do here. The basis for this is an account of addition and multiplication, but we shall here omit their derivation from the basic assumptions that determine the series of natural numbers, because this is common ground in foundational enquiries and can be taken as well-known.

Extensions of the domain of number are attended by a radically mistaken conception that brings with it disastrous consequences, namely in the view (already criticized in Section II) that in introducing new symbols we introduce something factually new, or 'create' new mathematical objects.

In fact, however, legitimate statements about these new numbers are simply statements about natural numbers, and operations with these new numbers are simply operations with natural numbers.

The primacy of natural numbers has indeed been repeatedly emphasized by the greatest mathematicians,[26] and we may even say that on this matter those who enquire into the foundations of mathematics today are on the whole agreed; but at two decisive points, namely the definition of limiting values and the definition of irrational numbers, the connections were often not grasped with sufficient clarity. In what follows we shall show how important a dissolution of the symbolism of rational and real numbers is for a correct view of the transfinite; moreover I am convinced that this move is apt to lead to 'internal mathematical' progress as well. Above all I am here thinking of a better understanding of the close connection between various mathematical disciplines that are mostly viewed as being largely independent from each other; but there are also other problems of higher analysis, as for example the criteria of convergence.

If, then, the whole of pure mathematics is contained in nucleus in the theory of natural numbers,[27] an unambiguous definition of the concept of natural number and of the 'operations' with natural numbers must yield the relation of mathematics to logic. We conclude this section by clarifying this connection.

In our analysis of logic we have observed that two aspects are essential for the concept of logic, namely the tautological character of logical propositions and the formal character of logical concepts.

Let us begin with the second criterion and note that it applies to mathematics as well; for, as we have recognized, the natural numbers fall into the formal domain. What of the tautological character of mathematical propositions?

Here the following distinction is important: mathematical propositions can be divided into two disjoined classes, the first containing those propositions that are finite (even in the sense of the prevailing view), that is, those in which we make no use of the fact that no number is the greatest number. Such propositions are tautologies: in the propositions $5 + 7 = 12$, for example, '$5 + 7$' means simply '12'.

Not so for the propositions of the second class, into which the presupposition of the unclosed character of the series of natural numbers does enter. For this presupposition is not required for defining particular natural numbers Z_n, where we need presuppose only all the numbers smaller than Z_n; thus no proposition asserting that for any number there always is a greater number or a greater number having certain 'properties', can be purely analytic. Nor can this be countered by the fact that such propositions, too, follow from the axioms of arithmetic. Here then we have reached a point at which we must introduce a separation.

The assumption that makes out this separation, namely that the number series is not closed, is factually justified by a consideration of the counting process in which there is no reason in principle why it should be stopped at any particular point.

Beside this distinction, which creates a division within mathematics itself, there is another important distinction that rests on the fact that natural numbers are formal singularities, while logic in the narrower sense does not contain formal singularities. Whether from this one is to conclude that the propositions of 'finite mathematics' should be terminologically separated from those of logic in the narrower sense and the name 'tautology' should be reserved for these latter, this is a question of heuristic expediency on which I do not wish to pronounce.

At any rate we recognize that we have said precious little if we answer the question whether mathematics is a part of logic with a brief 'yes' or 'no', without having precisely fixed the concept of 'logic'. Indeed, in dealing with foundation problems it is particularly important always to be fully aware how far the concepts arising in a question require further precision in order that we may regard the question as unambiguously determinate.

NOTES

[1] This cognitive fact was emphasized by E. Schröder, *Lehrbuch der Arithmetik und Algebra*, vol. 1, Leipzig 1873. Cf. also O. Stolz, *Vorlesungen uber allgemeine Arithmetik*, 1885, Part I, p. 9f.; L. Kronecker, 'Über den Zahlbegriff', *Werke*, 1899, vol. III, 1, p. 249ff.; H. Helmholtz, 'Zählen und Messen', *Wissenschaftliche Abhandlungen*, vol. III, p. 356ff.; O. Hölder, *Die Arithmetik in strenger Begründung*, 2nd edition, Berlin 1929, p. 14ff.

[2] Cf. also Burkamp, 'Begriff und Beziehung', l.c., p. 182ff.; E. Cassirer, *Philosophie der symbolischen Formen*, Part 3, Berlin 1929, p. 425ff.

[3] *Introduction to Mathematical Philosophy*, London 1919, p. 15.

[4] 'Philosophie der Mathematik und Naturwissenschaft', l.c., p. 28.

[5] For what follows, cf. also p. 53 above.

[6] The most important analyses figure in the *Critique of Pure Reason* in the chapter 'On the Schematism of the Pure Concepts of Reason'.

[7] Thus, for example, the definition of Kronecker (l.c., p. 256): "The number of objects is therefore a property of the assembly as such; that is, of the totality of objects conceived independently of any particular arrangement". An essentially similar view has recently appeared in Hilbert and Ackermann, *Grundzüge der theoretischen Logik*, 2nd edition, p. 109: "A number is not an object in the proper sense but a property. The individuals that possess a number as property cannot be the things counted themselves, since each of these is only one, so that a number different from one could never arise. However, we may conceive number as a property of that concept under which the chosen individuals are united. For example the fact that the number of continents is five cannot be expressed by saying that the number five belongs to each continent, but it is indeed a property of the predicate 'being a continent' that it applies to precisely five individuals."

That this view is wrong we have already established at the beginning of this section. It is particularly dangerous because it is a main motive for the introduction of the extended functional calculus, the focal point that forms logico-mathematical pseudo-problems.

[8] Frege defines as follows: the number that belongs to the concept F is the denotation

of the concept 'of the same number as the concept *F*'. *Die Grundlagen der Arithmetik*, Breslau 1884, p. 80. Cf. also Burkamp, l.c., p. 182ff.

[9] "The cardinal number of a class α . . . is defined as the class of all classes similar to α, . . ." *Principia Mathematica*, vol. II, p. 4. It is interesting to observe a certain similarity between Russell's view and that of J. S. Mill, who states (in a note to James Mill, *Analysis II*, p. 92): "Numbers are in the strictest sense names of objects. Two is certainly the name of things that are two, two spheres, two fingers and so on." On the other hand, Wittgenstein has cut himself free from Russell's theory of classes in mathematics. In *Tractatus*, 6.031, he says "The theory of classes is altogether superfluous in mathematics. This is connected with the fact that the generality needed in mathematics is not chance generality."

[10] This will become completely clear in the following analysis of complete induction.

[11] I owe an improvement of my original formulation to Carnap.

[12] The reader familiar with the axiom system of set theory may find that this formulation reminds him of Fraenkel's axiom of limitation (*Zehn Vorlesungen*, p. 102) by which he excludes those sets that are not implied by the axioms. Against this axiom J. von Neumann has directed his own critical remarks ('Eine Axiomatisierung der Mengenlehre', *Journal f. Math.* **154**, 219ff., esp. p. 229). The gist of this criticism is that an axiom of limitation can be consistently added to a system of axioms only if that system has categoricity, that is, if it is monomorphic. Now the monomorphism of progressions has been proved in *Principia Mathematica* (vol. III, p. 146, proposition 263.16); but our requirement of limitation, as we shall establish in what follows, is equivalent to the principle of complete induction; the three axioms along with that requirement thus determined progressions. Therefore if we contest the proof in *Principia Mathematica* (as Neuman does, according to an oral communication), then the objections made on this basis against our formulation relate equally to Peano's axiom system. We shall return to this below, p. 74.

Concerning related problems about 'completeness axioms' in analogy to the one used by Hilbert in his axioms of geometry, cf. also M. Geiger, l.c., p. 265ff.; P. Finsler, 'Über die Grundlegung der Mengenlehre', I, *Math. Zeitschr.* **25** (1926), 683ff.; R. Baer, 'Über ein Vollständigkeitsaxiom in der Mengenlehre', *Math. Zeitschr.* **27** (1928), 536ff.; R. Baldus, 'Zur Axiomatik der Geometrie I', 'Über Hilberts Vollständigkeitsaxiom', *Math. Ann.* **100** (1928), 321ff.; Fraenkel, who has kindly read the proofs of the present book, has pointed out to me that the replacement of Peono's axioms by a requirement of limitation essentially coinciding with that mentioned above was carried out by him in his article 'Axiomatische Begründung von Hensels p-adischen Zahlen', *Journal f. Math.* **141** (1912), 43–76, p. 49.

[13] 'Intuitionistische Betrachtungen über den Formalismus', *Sitzungsber. d. Preuss. Akad. d. Wissensch., Phys. math. Kl.* (1928), 48–52, p. 49.

[14] The most important passages are in the note on the thesis of the first antinomy. There Kant says: "The true (transcendental) concept of infinity is this: that the successive synthesis of unity by measuring a quantity can never be complete." The final sentences of the note deserve special emphasis: "Since this synthesis would have to amount to a series that can never be completed, we cannot think of a totality ahead of it nor yet by means of it. For the concept of totality itself is in this case the intuition of a completed synthesis of the parts, and this completion is impossible, as therefore also the concept of it."

[15] *Arithmetices principia nova methodo exposita*, Turin 1889.

[16] Russell has transformed this axiom system as follows (cf. *Principia Mathematica*, vol. II, § 122, p. 253ff., and *Introduction to Mathematical Philosophy* p. 7ff.) He uses the relation 'precursor of' (P) as the only basic concept. Its domain is fixed by the following stipulations: (1) P is a one-one relation. (2) P has just one initial term. (3) The whole domain of the relation is contained in the posterity of the initial term. (4) The relation has no final term.

[17] For this point and for the whole problem, cf. J. König, *Neue Grundlagen der Logik, Arithmetik und Mengenlehre*, p. 155ff.; O. Hölder, *Die mathematische Methode*, Berlin 1924, p. 331ff.

[18] We shall show later that such a series does not possess a sense analogous to that of the sequence of natural numbers, but we shall leave this open for the present in order not to pile up difficulties.

[19] Cf. *Science and Hypothesis*, p. 9. "The essential character of reasoning by recurrence is that it contains, condensed, so to speak, in a single formula, an infinite number of syllogisms." Also *Science and Method*, Book 2, Chapter 3. It must be emphasized that alongside this 'synthetic principle' Poincaré assumes the existence in mathematics of other similar principles. He remarks (*ibid.*, p. 149f.): "If a property holds for the number 1 and if one observes that it also holds for $n + 1$ provided that it holds for n, it will hold for all integers. It is in this that I saw the essence of the mathematical mode of inference. By this, I did not mean to say, as has been believed, that all mathematical modes of inference can be reduced to an application of this principle. If one examines them more closely, one will see applied in them many other analogous principles that exhibit the same essential properties. In this category of principles complete induction is merely the simplest of all, and for that reason only I used to describe it as typical."

In the years 1904 to 1909 a lively discussion on the problems of complete induction took place between Poincaré, Russell and Couturat in *Revue de Métaphysique et de Morale*. Poincaré's view has influenced neo-intuitionism and the latest formalist publications also come near to it. Cf. J. v. Neumann, 'Zur Hilbertschen Beweistheorie', l.c.

[20] This means that no change of place takes place as between axioms and derived theorems.

[21] However, this does not yet establish whether the mathematical description of the world is 'simple'.

[22] "Inductive cardinals are those that obey mathematical induction starting from 0, i.e. in the language of Part II, Section E, they are the posterity of 0 with respect to the relation of ν to $\nu +_c 1$ or, in more popular language, they are those that can be reached from 0 by successive additions of 1". *Principia Mathematica*, vol. II, p. 207.

[23] *Introduction to Mathematical Philosophy*, p. 27.

[24] 'Beiträge zur Begründung der transfiniten Mengenlehre', *Math. Ann.* **46** (1895), 481–512, p. 481.

[25] 'Über die Antinomien der Prinzipien der Mathematik', *Math. Zeitschr.* **14** (1922), 236–243; 'The Theory of Constructive Types (Principles of Logic and Mathematics)', Part I and II, extracted from the *Annales de la Société Polonaise de Mathématique*, Cracow 1923/25; and more recently 'Über die Hypothesen der Mengenlehre', *Math. Zeitschr.* **25** (1926), 439–473.

[26] Cf. Kronecker's well-known dictum "God made the integers, all the rest is the work

of man." In his lectures, Weierstrass, too, unambiguously expressed his conviction of the primacy of the natural numbers. Husserl (*Philosophie der Arithmetik*, Leipzig 1891, p. 5) quotes the following from Weierstrass' lectures (summer 1878, winter 1880/81): "Pure arithmetic (or pure analysis) is a science that is based exclusively on the concept of number. For the rest, it requires no presuppositions, postulates or premisses of any kind."

For the development of the conception of number in the 19th century, cf. G. Stammler, *Der Zahlbegriff seit Gauss*, Halle a. S. 1925.

[27] That this also holds for the pure geometries will become completely clear in Section IV.

IV

NEGATIVE NUMBERS, FRACTIONS AND IRRATIONAL NUMBERS

The simplest path towards understanding the so-called extensions of the number concept lies through the operations inverse to addition, multiplication and potentiation. Let us head our investigations with an observation by Russell that lays bare the basic mistake in the ingrained conception of these new 'numbers': "One of the mistakes that have delayed the discovery of correct definitions in this region is the common idea that each extension of number included the previous sorts as special cases. It was thought that, in dealing with positive and negative integers, the positive integers might be identified with the original signless integers. Again it was thought that a fraction whose denominator is 1 may be identified with the natural number which is its numerator. And the irrational numbers, such as the square root of 2, were supposed to find their place among rational fractions, as being greater than some and less than the others, so that rational and irrational numbers could be taken together as one class, called 'real numbers'. And when the idea of number was further extended so as to include 'complex' numbers, i.e. numbers involving the square root of -1, it was thought that real numbers could be regarded as those among complex numbers in which the imaginary part (i.e. the part which was a multiple of the square root of -1) was zero. All these suppositions were erroneous, and must be discarded ... if correct definitions are to be given."[1]

Let us begin with subtraction and, as in Section III, proceed from an exhibition of the temporal model, and only afterwards form the logical abstraction that corresponds to this model.

The meaning of subtraction is seen when we consider the decomposition of the counting process into partial processes. If in the 'successive synthesis of the manifold' one has reached let us say the fifteenth object or act of apprehension, one can ask oneself how many objects have been apprehended after the tenth counted object, or how many between the third and the ninth. In this way one obtains the definition of subtraction as the inverse operation to addition. However, this definition is tied to the presupposition that the number subtracted is not bigger[2] than the number from which it is subtracted; but the seeming extension of this operation beyond this boundary is an abbreviated mode of expression for a train of thought of a different kind. For we are here concerned with the relation between two 'oppositely directed' domains, that is, with the determination of a resultant by subtraction of two components of the same kind, where, to begin with, it appears undetermined which of the two is to be subtracted and from which it is to be subtracted. Simple examples are: the resultant of two forces acting in opposite directions, or the balance of a bilateral credit and debit relation. If one speaks of a negative force, this means that the force acts in a direction opposite to the initial direction; if we speak of a negative credit of A with B, this means that B has greater credit with A (the 'initial person') than A has with B.

What makes the symbolism of negative numbers applicable to any relation is therefore the opposition of direction,[3] that is, the possibility of forming the resultant by means of subtraction.

From this the meaning of negative numbers emerges as follows: an element is called the $-n$th relative to a given counting process, if starting the count from it an element that in the given process is the mth now becomes the $(n + m)$th. We thus have a superposition of two counting processes, the second retaining the symbols fixed in the first, although their meaning has shifted.

The technical simplifications attained by operating with the symbolism of negative numbers become especially evident when a definite fixed point (origin) seems particularly suitable as a starting point for additions and subtractions. An example of such a 'natural' zero point we may mention is the one for pecuniary assets and debts of equal amount. The mode of calculating with negative numbers and especially the dissolution of the seemingly paradoxical formation of products with a negative number as multiplier is readily intelligible after these reflections. That 'plus times minus' yields 'minus', and 'minus times minus' 'plus'[4], simply means the making absolute of the laws of subtraction, according to which $(a - b)(c - d) = ac - bc - ad + bd$. The negative number is thus a subtrahendum (number and operational sign) isolated from the framework of subtraction.[5]

We can thus indicate no common logical property of negative numbers on the basis of which they might be separated from natural numbers.

The case is somewhat more difficult when we try to understand the principles that underlie the formation of fractions. Here we start from the operation inverse to multiplication. It is called 'division' and what characterizes it is that the product and one factor are given while we are to find the second factor. Since multiplication obeys the commutative law there is no need to distinguish between multiplicandum and multiplier.

It is immediately obvious that on this definition we can attach no meaning to a division that 'leaves a remainder'. For evidently there can be no further element between the nth and $(n + 1)$th.

Closer investigation then shows without difficulty that the symbols called 'fractions' are not an extension of number, but symbols for relations between natural numbers.[6] Indeed, 'for fractions' one stipulates by definition that for two fractions p/q and r/s we have $p/q \lesseqqgtr r/s$ if $ps \lesseqqgtr qr$. Further, the basic operations of 'addition' and 'multiplication' 'for fractions' are defined in such a way that the commutative, associative and distributive laws hold.

All operations carried out with fractions are then in fact operations with natural numbers that have been given another name. For example the sum of two fractions is determined by forming the sum of the products 'numerator of the first fraction times denominator of the second fraction' and 'denominator of the first fraction times numerator of the second fraction' and setting it down as numerator of the new fraction, its denominator being the product of the denominators of the first two fractions. All mathematical connections performed in this procedure are operations with natural numbers: the introduction of 'fractions' is thus effected by nominal definitions 'with the help of the natural numbers'; that is, it amounts to a prescription concerning the use of certain symbols, which does not introduce any new logical aspects. It is therefore not, as assumed on the basis of uncritical views, that there are both natural numbers and fractional numbers given 'from the outset' as independent domains, between which we 'subsequently' discover relations, but a fraction is simply an incomplete symbol, that is, part of a symbol for certain operations with natural numbers.

How, then, can we explain that fractions are nevertheless mostly regarded as independent 'rational numbers' alongside natural numbers, and that in the end even these last have been viewed as special rational numbers with the denominator 1?

The main reason for this lies in a faulty interpretation of the geometrical facts of measurement, in which one imagines one is confronted with fractions, intuitively, as it were.

By measuring we understand the fixing of covering relations for spatial structures with the help of counting processes.

A distance f is measured by means of the 'unit distance' u, by ascertaining how often u can be laid on f. If the nth u exactly covers f (that is without sticking out beyond the end), then we assign the measure n to the distance f.

What happens if an exact covering of this kind of f by u is not realizable? Then, because we here assume that the axiom of Archimedes holds, there is a smallest number n, so that f can be completely covered by the nth u, only this time that u will protrude beyond f. This state of affairs can also be symbolized by the formula $(n - 1) u < f < nu$. In order more closely to determine the remainder r of f, which lies between the part to which we have assigned the number $n - 1$ and the end point of f, we use the following procedure. We adopt a distance u_1 as a new unit (measuring standard), such that u_1 is smaller than u,[7] not greater than r and so chosen that u can be exactly covered by it (say, p times); that is, no part of the last u_1, used will protrude beyond the end point of u. It is readily seen that these three conditions can always be realized together. Having chosen such a u_1 it may happen that the remainder r can likewise be exactly covered by u_1, let us say by q successive applications. Then, in view of the third condition, the distance f can also be covered exactly by u_1, the endpoint of f will coincide with that of $(p(n - 1) + q)$-th u_1. With the help of these reflections we specify the 'addition of distances' and the 'multiplication of a distance with a natural number', which automatically also defines the corresponding inverse operations of subtraction and division for the domain of rational numbers. Thus, the equation $u/f = 1/n$ is merely a different form of the equation $nu = f$, where u is called the nth part of f.

The seeming intuitive reality of fractions in measurement then rests on a misinterpretation of the presupposition that there are no smallest and no biggest distances in space, so that the unit of measurement can be assumed to be arbitrarily small. If we first choose a unit u, the application of the symbolism of rational numbers to the measurement of a distance u_1 that can be laid on u exactly five times will mean that we have to assign to u_1 the fraction 1/5; but what is intuitively given by this fact is at best[8] the distance, and not the fraction assigned to it. Since direction in space implies an ordered structure and spatial covering contains the structure of equality, by combining these two aspects we can engage in metrical geometry; but we must not imagine that by means of geometrical intuition we can logically justify an extension of the number concept.

This raises the question of principle concerning the character of geometrical intuition, which we must briefly analyse, since it plays an important part in the problems of the infinite in mathematics.

This analysis will have to clarify the peculiar double position that intuition occupies in geometry insofar as on the one hand it is regarded as a source of knowledge, as a 'last' intuition, neither capable of nor in need of further theoretical justification, which scientific thought must simply accept and arrange into a systematic order, while on the other hand it is set over against exact thought as a 'vague' intuition that should not be trusted. There are here two aspects that need to be considered separately: firstly the relation between intuition and thought, and secondly the alleged intuitive character of geometry.

As regards the first point, we must at once insist that to treat intuition and thought as opposites is incorrect, though it is often done. We can effect a distinction that captures the opposition intended only within the domain of thought, namely between intuitive and non-intuitive thinking; by the latter we are to understand formal thinking that relates to the structure of the world disregarding the sense-dependent content.

The vagueness of intuition is however mostly urged against such mental processes as lead to premature extensions of certain simple intuitive findings, which can involve erroneous exclusions of actual more complicated formations. Such inadmissible generalizations are by no means confined to spatial intuition but form a main source of mistakes of thinking in general. In special cases a critical going back to the arithmetical structure of intuition and mentally operating with this structure not infrequently leads to a removal of such errors, because retranslating structural relations into intuitional form (model formation) can discover new objects of intuition that refute the original false assumption that such objects are impossible. An example is Peano's curve,[9] which contains all points of a surface and thereby contradicts the assumption that a multidimensional continuous spatial structure contains 'more' points than a one-dimensional such structure.

Even more dangerous than severing intuition from thinking is the erroneous assumption that certain number statements operating with the 'infinitely big' or the 'infinitely small' can be intuitively realized, this seeming intuitive presentation being regarded as rendering the operations legitimate. That such a presentation is in fact not possible is to be traced back not to a 'defect of intuition' but to incorrectness in the arithmetical formulations. However, these latter enter into the very principles of geometry or of the various geometries, and here we come to our second point; for these geometries are simply analyses of spatial structures with the help of systems of relations, as we have noted earlier and shall further clarify below.

If now we are to make the character of geometrical propositions completely plain and grasp what is their cognitive content, we must first acknowledge that it is erroneous to assume that geometrical concepts are concepts of intuitive objects: they are on the contrary abstractions, of a kind to be specified presently; they are 'idealizations' of intuitive data.

What is intuitively given is only extended corporeal things, but not the surfaces, curves and points of geometry, which is why we must always go back to what is given in three dimensions if we are to find a model that is genuinely intuitive. For instance, to the proposition that two intersecting surfaces have a curve as a common element corresponds the fact that if two interpenetrating bodies are made progressively smaller in a given dimension this produces a boundless diminution of the common element towards two dimensions. Quite generally, a two-dimensional structure cannot really be understood except as a body conceived as subjected to an arbitrarily far-reaching diminution in one dimension. One-dimensional structures (curves) are arbitrarily small in two dimensions and 'non-dimensional' points in all three.[10]

Arbitrary smallness, however, does not mean that we exclude extension, but merely that we exclude a fixed lower bound to it.[11]

Accordingly the thesis that curves can be made up of points, surfaces of curves and bodies of surfaces, will yield intuitive sense only if we regard a point as an arbitrarily small piece of curve, for the view that infinitely many lower-dimensioned structures can be intuitively assembled into a higher-dimensional structure is a non-thought.

These observations about the 'infinitely small' can be analogously applied to the 'infinitely large' in geometry. 'Infinitely long' curves and surfaces 'infinite in two dimensions' are merely abbreviated expressions for excluding fixed upper bounds to extension. If we are clear about this, we may use the well-known and convenient mode of expression that has become established amongst mathematicians; but it would be welcome if their meaning were to be made plain from the start in textbooks of geometry.

The above explanations about geometrical structures already show up one important non-intuitive aspect within geometrical cognition: for, evidently, that there is no lower or upper bound to extension cannot be intuitively grasped. It is particularly important to be clear on this, because all attempts at intuitively verifying transfinite statements of arithmetic and set theory operate essentially with the infinitely small and infinitely large in geometry, that is, with the arbitrariness of the upper or lower bound to extension. As we have already recognized, the position is rather that some non-intuitive aspects enter the basic concepts of geometry themselves, namely precisely those aspects from which the misinterpretations about the transfinite start in arithmetic (analysis). Since these misinterpretations ('infinitely small' for 'arbitrarily small' and 'infinitely large' for 'arbitrarily large') begin even at the level of the basic concepts of geometry, the illusion arises that the transfinite in spatial structures (the object of geometry) is intuitively given. To this fascination of the seeming intuitive reality of the transfinite even front rank mathematicians and philosophers have succumbed,[12] and it is hardly to be assumed that without this prejudice Georg Cantor's doctrine of the actual infinite of various powers could have established itself. (To Cantor's theory we shall return in Section V.)

If now we ask for the epistemological position of the various geometries in the system of sciences, which is determined by their relations to arithmetic on the one hand and to physics on the other, we must distinguish the following three layers that are usually not sufficiently kept apart in philosophical discussions about the nature of geometry; the correct analysis will then already constitute the answer to these questions.[13]

(1) The formal nucleus, which in modern axiom systems of geometry constitutes the almost exclusive object of discussion.[14] It is a formal system of logico-arithmetical relations and therefore the same in structure as certain partial domains of the domain of arithmetic, which latter is the universal schema of formal relations. How far we shall require that these structures be shared by intuitive space, in order to call a system of relations by the name 'geometry' is a matter of convention; today we distinguish, alongside non-Euclidean geometries, non-Archimedean and non-Cartesian geometries, and we certainly do not confine ourselves to systems of only three parameters.

(2) The intuitive (or, better, quasi-intuitive) spatial model. This amounts to an 'idealization' of data of perception in the sense explained above. As such it can never be confirmed or refuted by empirical findings. If different models have the same formal nucleus, they are called isomorphic.

(3) Empirical validity. This depends on how suitable a system of relations is for formulating natural connections and is therefore largely determined by empirical facts. We know that classical mechanics uses the 'language' of Euclidean geometry, and the general theory of relativity that of Riemannian geometry.

Let us now return to the symbolism of fractional numbers. The mathematically most important concept connected with it is that of a limiting value of a sequence of rational numbers. This is precisely the point where mathematicians who operate with this symbolism not infrequently stand in danger of ceasing to see clearly that this involves ascertaining relations between natural numbers.

We call r/s the limiting value of the sequence $p_1/q_1, p_2/q_2 \ldots, p_n/q_n \ldots$ if for an arbitrarily small positive fraction h we can find a number n so that for the nth term of the sequence and for every subsequent term the following inequality holds: $|r/s - p_n/q_n| < h$.[15] If we now determine the meaning of this concept without using the symbolism of fractions, used in the definiens, we obtain the following definition: given two sequences of natural numbers $p_1, p_2, \ldots, p_n, \ldots$ and $q_1, q_2, \ldots, q_n, \ldots$, then r/s is the limiting value of $p_1/q_1, p_2/q_2, \ldots, p_n/q_n, \ldots$ if for every natural number k a natural number z can be found such that for every natural number $n \geq z$ the relation $q_n s > k|q_n r - p_n s|$ holds.

In analysis the concept of a limiting value is dealt with in connection with that of a bounded sequence. Since the latter concept too is important for the following investigations about irrational numbers, let us include it in our examination.

A bounded sequence, in the sense of analysis, is one all of whose terms (presupposed to be greater than zero) lie below a certain number. We see at once that after dissolution of the symbolism of rational numbers the sense of this definition is not that all elements of a sequence of natural numbers lie below a certain natural number; what it does mean is that for every element of the sequence there is an upper limit that varies with the element's ordinal index within the sequence. After this preliminary remark the following 'translation' of the concept of a bounded sequence will be completely obvious.

Given two sequences of natural numbers $p_1, p_2, \ldots, p_n, \ldots$, and $q_1, q_2, \ldots, q_n, \ldots$, then $p_1/q_1, p_2/q_2, \ldots, p_n/q_n, \ldots$ is called a bounded sequence if for a fixed natural number i and arbitrary n the relation $iq_n > p_n$ holds.

It is easy to show that only bounded sequences have a limiting value. On this point there is thus no divergence from the usual terminology established in mathematics.

However, analysis asserts that, for a monotonic[16] sequence, being bounded is not only necessary but also sufficient for its having a limiting value. We shall soon have to investigate whether this assumption is justified.

On the pattern given above, the remaining propositions of analysis about limiting values and about relations into which the concept of a limiting value enters (such as 'continuity') may now be 'translated'.

For example, the theorem of analysis that 'no sequence has more than one limiting value' can be 'translated' into 'for no sequence is there more than one pair of relatively prime natural numbers satisfying the condition for the limiting value just given'.

This example already shows clearly how much more simple it is to formulate such propositions in the usual way using the symbolism of rational numbers than if we deny ourselves this device. However, we must emphasize once again that it is not the case that there is 'first' a limiting value of a sequence of rational numbers which can then be replaced by relations between natural numbers, but that the operations actually performed in calculations with rational numbers are precisely those operations with natural numbers that we have described above. Indeed, as we have seen, calculating with fractions is simply calculating with the numerators and denominators of the fractions; it is only if we hypostatize newly introduced symbols into mathematical objects (a procedure already criticized above) that we are led to misinterpret fractions as objects of mathematical enquiry existing alongside and independently of natural numbers. Of course this analysis is not intended to criticize the usual symbolism of mathematics which technically is almost indispensable, but only the misinterpretation of that symbolism which leads to sham problems of the most questionable kind. Actually, these problems do not reveal themselves in their full danger until we reach the level of 'extension of the number concept', with the introduction of irrational numbers, to which we shall now turn. The roots of the difficulties that emerge here do however reach back in large measure to the semantic obscurities, just analysed, in the symbolism of rational numbers.

Since we have grasped its meaning in principle, we can in what follows safely operate with that symbolism, which will prevent a piling up of difficulties. We shall, however, be constantly aware that all meaningful statements about rational numbers are statements about natural numbers. Only at the most important points shall we explicitly indicate the underlying relations between natural numbers. Let us note briefly by the way that in contrast with conventional usage we have here been using the concepts 'fractions' and 'rational numbers' as synonyms; for the ordinary use of the term 'rational number', which covers positive and negative integers and fractions, has turned out to be a logical monstrosity.[17]

Let us begin by considering the two operations inverse to potentiation, namely the extracting of roots and the taking of logarithms.

The procedure of potentiation itself (raising to a power) as that of multiplying two or more equal factors needs no detailed consideration. Moreover, we readily see that to this procedure there correspond two different 'inverse' processes; namely first the finding of the base or root, that is, of the repeated factor, if its exponent (that is, the number of times it appears) and the whole expression are given; and secondly the finding of the exponent if the whole expression and the root are given. The first procedure is called the extraction of roots and the second the taking of logarithms. However, the operation of potentiation along with its inverse operations are defined only for positive integer exponents, as far as our above stipulations go, while the base can be any rational number. Nothing new enters when we introduce negative and fractional exponents, for this merely amounts to a short-hand symbolism not corresponding to any specific procedures. For we simply stipulate by definition that the operational sign $(1/a)^n$ is equivalent to a^{-n} and the operational sign $^n\sqrt{a^m}$ to $a^{m/n}$. Thus the equation $4^{-2} = 1/16$ simply says the same as $(1/4)^2 = 1/16$ and $4^{3/2} = 8$ the same as $^2\sqrt{4^3} = 8$. What, then, of the case where we can prove that no x satisfies the initial equations $x^b = p$, or $b^x = p$?

To start with, we must acknowledge this state of affairs without reservations. Extracting roots and taking logarithms are defined only as inverse operations of an 'actual' potentiation, that is, one derivable from the counting process; if such a potentiation cannot be provided, the task of extracting roots or taking logarithms would amount to specifying two operations inverse to an operation that did not exist, which is obviously absurd. No mathematician can by a mere decree impose sense on what is senseless, building up a 'free creation of the spirit' alongside reality; at best he can, with due caution, use the same symbolism for different operations. For example, as we shall see presently, the root sign in $\sqrt{2}$ means something quite different from what it means in $\sqrt{4}$, but this symbolism is of great heuristic convenience in view of certain problems.

In our case the extension of the symbolism is based on the following considerations: that the number 2 is not the square of a rational number is easily proved by a well-known argument. For suppose that 2 were the square of m/n (where we may assume that the natural numbers m and n are not both even, since otherwise we could cancel by 2 until one of the two numbers becomes odd); thus let $m^2 = 2n^2$. Hence m^2, the double of

n^2, would have to be even and therefore so would m itself, so that m^2 would be divisible by 4. Since by hypothesis m and n are not both even, n would have to be odd, so that $2n^2 = m^2$ would not be divisible by 4, which amounts to a contradiction. We can, however, indicate a procedure (called the extraction of the square root) that yields a sequence of rational numbers each of which is bigger than its precursor (and therefore bigger than all its precursors), where the squares of these numbers always remain smaller than 2, but approach it arbitrarily closely. If, therefore, we specify an arbitrarily small rational number μ, we can always find a rational number n, to be obtained from that procedure, such that $2 - n^2 < \mu$; that is, the squares of the rational numbers obtained in the procedure can approximate arbitrarily closely to the number 2.

The two requirements to which the extraction of roots is subject must be kept well apart; in the case of our example they are first, that the squares of rational numbers to be determined are to fall short of 2 by less than any arbitrarily small rational number given in advance; and secondly, that the squares must never exceed 2. The second condition is satisfied if for an arbitrarily small rational number μ we have the inequality $2 + \mu > z_n^2$, where z_n is any of the rational numbers obtained in the extraction procedure. To prove that this condition holds we indicate a second and descending sequence of rational numbers whose terms boundlessly approach 2 'from above' and stipulate a law of mutually one-one correlations of pairs between the elements of the two sequences such that every element of the second sequence is greater than the element from the first sequence paired with it. However, we must not interpret this state of affairs in such a way as to assert that 'between' these two sequences there is something else that is enclosed by them within arbitrarily narrow bounds, and this something could be defined as an irrational number. The fact that in the first of the two sequences indicated we cannot point to a greatest rational number must not be reinterpreted into the assertion that there is a limit (the irrational number), to which both these sequences approach indefinitely without ever being able to reach it. As to the sequence of rational numbers[18] the squares of whose partial sums or partial differences respectively come indefinitely close to the number 2 (converge to 2), it is equally incorrect to regard it as an independent totality and to define it as an irrational number; for we cannot form totalities having infinitely many terms.

It is thus also an incorrect mode of expression to assert that the rational numbers calculated in the procedure of extracting roots indefinitely approach $\sqrt{2}$,[19] for there is no such number, at least not at this stage of our inquiry; to speak correctly one might at most say that the squares of those rational numbers indefinitely approach the number 2.

If now we wish to ascertain what is the sense of this last assertion, once we have eliminated the symbolism of rational numbers, it is again enough to consider clearly what are the operations that are actually performed in extracting roots. We then recognize that the procedure for calculating $\sqrt{2}$ consists in successively working out the greatest natural numbers, whose squares are smaller than 2.10^0, 2.10^2, 2.10^4, 2.10^6 and so on.[20] In this way we obtain the numbers 1, 14, 141, 1414 and so on. Quite in general, we recognize that an 'infinite decimal fraction' simply means a sequence of natural numbers, where, as repeatedly observed above, we must understand by 'sequence' not an infinite totality, but the domain of a certain relation (law). If in our example we denote the numbers of that sequence by $z_0, z_1, z_2, \ldots, z_n, \ldots$, then the 'arbitrarily close approximation' of the squares of these numbers to the number 2 consist in the fact that for any number k we can find a number m such that for any $n \geq m$ we have $(2.10^{2n}/(2.10^{2n} - z_n^2)) > k$.

After this example we can proceed to formulate the general state of affairs that underlies the formation of the concept of irrational number.

To this end we shall start from the limit of a bounded sequence, this time defining it with the help of the symbolism of rational numbers in the sense of analysis.

Given a bounded sequence $F(f_1, f_2, \ldots f_n, \ldots)$, fixed by its law of formation, given further a rational number G, then let G be called the limit of F, if for every arbitrary rational number k we can find a natural number h such that for every $i > h$ we have $G - k < f_i < G + k$. We then say that F converges to G.

If the sequences are monotonic, this definition makes the limit either the smallest rational number greater than any number of the sequence (namely for increasing sequences), or the biggest rational number that is smaller than any number of the sequence (namely for decreasing sequences).

However, our example of the square root of 2 already shows that not every bounded sequence of rational numbers has a limit. For if, for a sequence of rational numbers whose squares differ from 2 by less than an arbitrarily small amount, there were a least number greater than any number of the sequence, then the square of that rational number would have to be exactly equal to 2. Yet we have proved that there is no such rational number. Still, as in the case of convergent sequences, here too the values come more and more closely together. The exact formulation of this state of affairs is contained in the following proposition:

In every bounded monotonic increasing (decreasing) sequence for any arbitrary rational number k we can determine an interval smaller than k, whose upper (lower) bound is a number that is greater (smaller) than any number of the sequence and which contains a number of the sequence.

We shall prove this proposition for increasing sequences; the proof for the decreasing case is to be conducted in corresponding manner.

Let the sequence in question be $f_1, f_2, \ldots, f_n, \ldots$ and the suitable interval smaller than k. Choose two rational numbers R_1 and R_2 such that R_2 is greater than any f_i while R_1 is not greater than any f_i. This is always possible, by definition (bounded sequence). If now $(R_2 - R_1) < k$, then (R_1R_2) is a suitable interval. Otherwise we determine an integer n such that $((R_2 - R_1)/n) < k$ and divide the interval (R_1R_2) into n equal sub-intervals; then either the last (highest) sub-interval contains an element of the sequence (in which case that sub-interval itself is a suitable interval), or amongst the n sub-intervals there is a lowest sub-interval whose lower boundary is bigger than any term of the sequence. This sub-interval cannot be the lowest of all of them, since according to our definition there must be a term of the sequence between R_1 and R_2; hence there is a next lower sub-interval and that will be of the suitable kind.

At this point, which is especially important for the problem of irrational numbers, let us now eliminate the symbolism of rational numbers, in order to bring out with complete precision the difference between rational and irrational 'limit'.

Given two sequences of natural numbers $p_1, p_2, \ldots, p_m, \ldots$, and $q_1, q_2, \ldots q_m, \ldots$, and a natural number h such that for any arbitrary natural number m we have the relation $hq_m > p_m$. Then for an arbitrarily large natural number k we can specify a number z and four other numbers r, s, t, u such that for every $n > z$ the following four conditions are fulfilled:

1. $ur > st$ (r/s is the upper bound and t/u the lower bound of the interval).
2. $p_nu > q_nt$;
3. $q_nr > p_ns$ (p_n/q_n lies within the interval).
4. $us > k(ur - st)$ (the interval is arbitrarily small).

That these conditions are different from those corresponding to the rational limit of a sequence is obvious on comparison.

The prevailing doctrine in analysis has however gone beyond the cognitive fact that there are bounded monotonic sequences of rational numbers that do not have a limit: for every such sequence, analysis postulates a limit.[21] For this purpose, the preferred procedure is the Dedekind cut in the domain of rational numbers.[22]

Such a 'cut' is defined as follows: if a set M of numbers is divided into two sub-sets M_1 and M_2 in such a way that every number belongs to one and only one of M_1 and M_2, each of M_1 and M_2 contains at least one number and every number in M_1 is smaller than every number in M_2, then this arrangement is called a cut in the set M. If M_1 has a greatest number P_1 or M_2 a least number P_2, then we say of P_1 (or P_2) that it 'generates' the cut. There are then four possibilities:

1. M_1 has a greatest number and M_2 a smallest number.
2. M_1 has a greatest number, but M_2 has no smallest number.
3. M_1 has no greatest number, but M_2 has a smallest number.
4. M_1 has no greatest number and M_2 has no smallest number.

In the first case the 'cut' is called a 'jump' in M, and in the fourth case a 'gap' in M, while in the second and third we speak of a 'continuous cut'. Dedekind then defines irrational number as a number that generates a cut that is a gap.

Such a cut without a point of application seems at first nonsensical. Still we might be tempted to save Dedekind's definition in principle by suggesting that what he had in mind were cuts defined by functions. Even then, however, an irrational cut is unacceptable. Let us illustrate this in the example of monotonic functions.

Given a rational number R and a monotonic function F, let M_1 contain every number z_i for which $R < F(z_i)$ and M_2 every Z_k for which $R \geq F(Z_k)$. Then M_2 will have a smallest element if and only if amongst the Z_k there is a number Z_{k1} such that $F(Z_{k1}) = R$. If there is no such number, then there is no rational number to generate the cut. If we are still to speak of a 'generation' at all, the cut is here produced by two inequalities. In this case the expression of a 'gap' is unsuitable too, for it seems to point to the possibility that a number could be left out or inserted, which does not apply. It is thus not possible to obtain irrational numbers as new entities, by means of the procedure of a cut in the domain of rational numbers.

If of course we accept Dedekind's definition, it is easy to prove the theorem, called after Weierstrass, that every bounded sequence has a cluster point (every bounded monotonic sequence a limit point).

Just as inadmissible as the definition of irrational numbers by means of a cut in the domain of rational numbers is the definition of irrational number identifying it with a bounded sequence.[23] On the contrary, we calculate with 'rational approximations', being aware that we can estimate and arbitrarily improve them. The definition of a concept, however, is correct only if it renders the meaning that the concept has in use, as we observed in Section I.

Finally, Russell's theory of irrational numbers is likewise untenable; for while, as we saw, it avoids the mistake of postulating limits to sequences where limits do not exist, it nevertheless makes essential use of the principle of comprehension, like Cantor's definition. Russell's train of thought is this: every Dedekind cut can be fixed by means of its lower class. Consider now a cut whose lower class has no maximum and call such a lower class a 'segment'. "Then those segments that correspond to ratios are those that consist of all ratios less than the ratio they correspond to, which is their boundary; while those that represent irrationals are those that have no boundary." Hence the definition: "An 'irrational number' is a segment of the series of ratios which has no boundary."[24]

We recognize at once that this definition of Russell's invites the same objections that we have raised against his definition of natural numbers; for on the present definition irrational numbers are classes of rational numbers. The consequences of this for Russell's system of mathematics we shall point out later.[25]

An 'irrational number' is then merely the abbreviated expression for the fact that to bounded sequences that have no limit there belong arbitrarily small clustering intervals.[26] Hence two bounded monotonic sequences without limit determine the same irrational number if every clustering interval of one of the sequences partially coincides with every clustering interval of the other (that is, the intervals share at least one rational number.)[27] It then follows without difficulty that the existing rules for calculating with irrational numbers are valid.

If the insight just gained is to be given expression in the formulation of mathematical propositions, we must inevitably formulate important theorems of algebra and analysis in a more complicated manner. For example, the 'irrational roots' of an algebraic equation do not actually satisfy it, but are merely a signification of the fact that there exists a law of formation for arbitrarily close approximations to satisfaction.[28] However, this evidently does not involve any factual change; it is merely a matter of calling things by the right name. As to the practical mathematicians' query what is the value of an investigation that results merely in complicating a well-tried simple terminology, our answer is this: a neglect of factual distinctions in terminology and symbolism is admissible only if there is clearly no danger that in our thinking we thereby blur those distinctions. Precisely when a mathematician in all good faith wishes to make use of these linguistic and semiotic abbreviations, he must clearly take into account what his words and signs mean in every particular instance, otherwise his own formation of concepts and stipulation of signs will give rise to sham problems of the most dangerous kind. This has come out with especial clarity in the case of the concept of irrational number, or the concept of real number comprising the rational and irrational numbers. More recently these sham problems have positively raised a wall preventing insight into the true character of the relations that were regarded as transfinite, because it was believed that irrational numbers could not be represented except by means of an infinite set,[29] so that the concept of an infinite set had to be viewed as indispensable for analysis. This in turn led to the view that the absurdities connected with the concept of the non-denumerable infinite constituted an upheaval of the foundations of analysis, since it seemed impossible to develop analysis without that concept. In fact, however, the non-denumerable infinite is out of place in analysis; the illusion to the contrary arises merely from misinterpretations.

We can easily see that no reformulation of the propositions of analysis on the basis of the insights obtained can alter anything in the stock of knowledge of that branch of mathematics, for this insight consists just in correctly describing the mathematical meaning of the symbolism used. The most important terminological differences arise from a change in Weierstrass' theorem on the cluster-point of a bounded sequence. As we have recognized, not every bounded sequence of rational numbers determines a number as its cluster-point, but every such sequence does determine arbitrarily small cluster intervals, which sometimes, but not always, contract to one number. However, as we shall see presently, there is no call for forming classes or totalities of such cluster intervals, and in this way the absurdities connected with the concept of an 'arbitrary real number' or a 'totality of all real numbers' simply disappear.

If we keep this in mind we can confidently use the symbolism of irrational numbers (which answers admirably to the technical demands of mathematics), without danger of falling prey to transfinite misinterpretations. These latter we shall examine more closely in Section V.

Let us now show that, with a clear understanding of the concept of irrational numbers, the absurdities arising with 'higher level irrational numbers' will likewise vanish: for we can eliminate them.

We will show this for the following propositions which lies at the centre of this complex of problems: 'every monotonic bounded sequence of irrational numbers has a limit'. If we eliminate the concept of irrational number, this proposition takes the form: for every monotonic bounded sequence of monotonic bounded sequences of rational numbers we can find an arbitrarily small interval that contains cluster intervals of all these sequences after a certain sequence (that is, for 'almost all sequences'). We shall give a proof for increasing sequences, which can be generalized without difficulty.

Given a bounded monotonic increasing sequence $f_1, f_2, \ldots, f_n, \ldots$ of monotonic increasing bounded sequences of rational numbers, we can find for an arbitrarily small rational number k a rational interval $I_k < k$ such that the following relation holds: let $z_1, z_2, \ldots, z_n, \ldots$ be a sequence of rational numbers with z_1 greater than any element of f_1, but not greater than every element of f_2, and in general z_n greater than every element of f_n but not greater than every element of f_{n+1}; this sequence is monotonic increasing, since $f_1, f_2, \ldots, f_n, \ldots$ is, and bounded, since by definition we can find for any of its elements a greater element in one of the bounded sequences $f_1, f_2, \ldots, f_n, \ldots$[31]. Then $z_1, z_2, \ldots, z_n, \ldots$ has arbitrarily small cluster intervals I_k, in the sense of the theorem proved above on p. 103. Now if such an I_k contains every element of $z_1, z_2, \ldots, z_n, \ldots$, starting with a certain element z_q, then it also contains a cluster interval of the sequence f_{q+1} and of all later sequences. For since z_q is not greater than every element of f_{q+1}, but z_{q+1} is greater than every element of f_{q+1}, an interval that contains z_q and z_{q+1} must contain a cluster interval of f_{q+1}, and generally an interval that contains z_{q+i} and z_{q+i+1} must contain a cluster interval of the sequence f_{q+i+1}.

This establishes our theorem and at the same time reduces the theorem about sequences of irrational numbers to a theorem about rational numbers, by which procedure we can eliminate irrational numbers of higher level. In this way, too, the vicious circle in the foundation of analysis disappears.[32]

Let us finally examine what becomes of propositions concerning the irrational roots of equations, once we dissolve the symbolism of irrational numbers. For the sake of simplicity we shall confine ourselves to algebraic equations with the further restriction that the coefficient of the highest power of the unknown shall be 1 and all the other coefficients and the independent terms are to be natural numbers.[33] These restrictions are, however, not important in principle; the general cognitive content that matters here will emerge clearly nevertheless.

Let then $x^n + a_1x^{n-1} + a_2x^{n-2} + \ldots + a_{n-1}x + a_n = 0$ be an equation possessing a positive[34] irrational root $p + q$, where p is the greatest natural number smaller than the root, namely the number before the decimal point, while q is the proper infinite decimal fraction after the decimal point.[35] Let the decimal fraction breaking off with the first, second, ... nth place be denoted by $q_1, q_2, \ldots q_n$. We will now represent this mathematical state of affairs without using the symbolism of irrational and rational numbers.[36]

Let there be given a sequence (f) of natural numbers $f_1, f_2, \ldots f_n, \ldots$ of the following form:

$$f_1 = r, f_2 = 10f_1 + s, \ldots, f_n = 10f_{n-1} + v, \ldots$$

where $f_1 = 10q_1, f_2 = 10^2q_2, \ldots, f_n = 10^nq_n$ and every $r, s, \ldots v$ signifies one of the numbers 0, 1, ... 9. Then for every natural number k however great we can find a natural number h such that for every natural number $i \geq h$ the following relations holds: if in the equation $x^n + 10^ia_1x^{n-1} + 10^{2i}a_2x^{n-2} + \ldots + 10^{(n-1)i}a_{n-1}x + 10^{ni}a_n = 0$ we replace x by the number $10^ip + f_i$, and the numerical value obtained for the polynomial on the left hand side (which may be a positive or a negative integer) is w, then the following relation holds: $((10^ip + f_i)/|w|) > k$.[37]

In contrast, the assertion that a natural number z is a root of the equation $x^n + a_1x^{n-1} + a_2x^{n-2} + \ldots + a_{n-1}x + a_n = 0$ states that z inserted into the left hand polynomial will give it the value 0.

We see that here two different mathematical states of affairs are denoted by the same name. Whatever the technical and heuristic advantages of the traditional terminology, and however foolish we should therefore be to give it up, none of this makes the two different states of affairs any the more alike. On the other hand, if we do not use that terminology and the attendant symbolism, 'irrational roots' are not somehow lost, only now they are differently named.

In conclusion we may mention the possible objection, that irrational numbers can be intuitively represented by geometrical constructions, let us say $\sqrt{2}$ by the diagonal of a square with side 1, or π by the circumference of a circle with diameter 1. Since, however, our general observations concerning geometrical intuition already suffice to show this objection to be invalid, no further comment is needed.[38] From these observations it follows that the incommensurability of two distances is in principle not verifiable by intuition. Even the assumption of a linear continuum falls away, once we clearly see that the possibility of 'intuitively' building up distances (curves) from points is merely an incorrect formulation of the cognitive fact that we can build them up from arbitrarily small distances (curves).

This removes the main prejudices concerning the transfinite, insofar as they arise from misunderstood interpretations of intuition and of the mathematical symbolism of rational and irrational numbers. Analysing the extension of the number concept that leads to imaginary and complex numbers would not add anything new in principle here, so that we can refrain from it.[39] In the next section we shall proceed to the core of the theory of the transfinite.

NOTES

[1] *Introduction to Mathematical Philosophy*, p. 63f.

[2] If minuend and subtrahend are equal, so that subtraction leaves nothing, we speak of the difference 0, but this difference cannot meaningfully figure as a factor in a multiplication. Thus the formation of the products $a \times 0$ or $0 \times a$ seems at first just as meaningless as that of the quotients: $a:0$ or $0:0$. However, we can understand multiplication by 0 as multiplication by an arbitrarily small number, so that 0 is then the concept of a limit. This will become fully intelligible after we have discussed rational numbers amongst which there is no smallest number. Similar considerations hold for 0 as exponent, in a^0.

[3] Gauss was already fully aware of this. He emphasizes explicitly that negative numbers may be used only "where what is counted has something opposite which when conceived as united with it must be equated with annihilation." Quoted in German by O. Becker, "Mathematische Existenz", l.c., p. 477.

[4] As regards the latter rule, the mathematician Clavius wrote in 1612: "it seems that we must blame the weakness of human intelligence for our inability to grasp by what stipulation this can be true". Quoted in Latin by Weyl, *Philosophie der Mathematik und Naturwissenschaft*, p. 26.

[5] A simple way of eliminating the symbolism of negative numbers is given by Kronecker (l.c., p. 345): "The concept of negative number can be avoided if in the formulae we replace the factor -1 by an unknown x, and the sign of equality by Gauss' sign for congruence modulo $(x + 1)$. Thus the equation $7 - 9 = 3 - 5$ is transformed into the congruence $7 + 9 \equiv 3 + 5x \pmod{(x + 1)}$; . . ." In the same work Kronecker likewise eliminates the symbolism of fractions and of algebraic numbers.

[6] Cf. the account in Euclid's *Elements* which goes back to Eudoxus; see also H. Hasse and H. Scholz, 'Die Grundlagenkrisis der griechischen Mathematik', *Kant-Studien* **33** (1928), 4–34; H. Scholz, 'Warum haben die Griechen die Irrationalzahlen nicht aufgebaut?', *ibid.*, pp. 35–72. In modern mathematical theory rational numbers are usually defined as ordered pairs of natural numbers.

[7] That is, we cannot entirely cover u with a single u_1.

[8] In this connection note the explanations immediately following.

[9] G. Peano, 'Sur une courbe qui remplit toute une aire plane', *Math. Ann.* (1890), **36** 157–160.

[10] Towards the end of the 19th century, Cantor's theorem that the points of an *n*-dimensional continuum could be put in one-one correspondence with those of a one-dimensional continuum and the construction of the Peano curve just mentioned seemed to endanger the exact mathematical definability of the concept of 'dimensions'. In 1911, however, Brouwer ('Beweis der Invarianz der Dimensionszahl', *Math. Ann.* **70**, 161–165) proved that there can be one-one continuous correspondences between Cartesian spaces of different dimensions. The precise definition of the concept of dimension was provided independently by K. Menger (cf. 'Zur Dimensions- und Kurventheorie', *Monatshefte f. Math. u. Phys.* **36**, 411ff.) and P. Urysohn ('Les multiplicités cantoriennes', *Comptes Rendus* **175** (1922), 440ff.). A comprehensive account of the theory of dimensions is given in Menger, *Dimensionstheorie*, Berlin 1928.

[11] Our considerations are thus completely different from those that Hjelmslev presents in his 'natural geometry' (geometry of approximation), since he expressly takes into account the empirical bounds to the intuition of the senses. Cf. Hjelmslev, 'Die natürliche Geometrie', *Abh. a. d. Math. Sem. d. Hamb. Univ.* **2** (1923), 1–36. However, there is a certain affinity with the views of M. Pasch. Cf. amongst other things his *Vorlesungen über neuere Geometrie*, Leipzig, 1882, and his essay 'Grundfragen der Geometrie', *Journ. f. Math.* **147** (1917), 184–190.

[12] Gauss on the contrary clearly understood that arithmetical assertions could not be verified by geometrical intuition, witness the following quotation: "It is my deepest conviction that the doctrine of space has quite a different position with regard to our knowledge *a priori* from the pure theory of magnitude; our knowledge of the former entirely lacks that complete conviction of its necessity (and therefore also of its absolute truth) which is peculiar to the latter; we must humbly admit that if number is only the product of our intelligence, space has a reality outside it as well: to this reality we cannot completely prescribe its laws in an *a priori* manner." (*Correspondence between Gauss and Bessel*, Leipzig 1880, p. 497ff., quoted in German by Kronecker 'Über den Zahlbegriff', l.c., p. 253).

[13] Cf. A. Einstein, *Geometrie und Erfahrung*, Berlin 1921; R. Carnap, 'Der Raum, ein Beitrag zur Wissenschaftslehre', supplementary fascicules to *Kant-Studien*, no. 56, Berlin 1922.

[14] This is the way Hilbert defines: "We think three different systems of things: the things of the first system we call points, denoting them by $A, B, C, \ldots$; the things of the second system we call straight lines, denoting them by $a, b, c, \ldots$; the things of the third system we call planes, denoting them by $\alpha, \beta, \gamma, \ldots$;" "We think of points, straight lines and planes as mutually related in certain ways, denoting these relations by words like, 'to lie', 'between', 'parallel', 'congruent', 'continuous', the precise and complete description of these relations is afforded by the axioms of geometry". 'Grundlagen der Geometrie', 3rd edition, *Wissenschaft und Hypothese* **7** Leipzig und Berlin 1923.

[15] The two vertical bars mean that we are to take the absolute value of the expression between them.

[16] A sequence of numbers is called monotonic increasing (or decreasing) if no term is greater (or less) than the next.

[17] Cf. Russell's dictum at the beginning of this section.

[18] In our present case these rational numbers are 1, 0.4, 0.01, 0.004, 0.0002 . . . and 2, 0.5, 0.08, 0.005, 0.0007 . . .

[19] The concepts of 'procedure' and 'approach' are to be taken quite timelessly. The 'procedure' consists in a one-one correlation between rational numbers and natural numbers (determination of the first *n* places). The fact that in actually calculating we proceed from one place to the next does not touch the sense of the principle by which the correlation is fixed 'simultaneously' for arbitrary *n*.

[20] The fact that powers of the number 10 enter at this point is of no arithmetical importance, since it is merely 'chance' that has led to the decimal system being introduced; but quite apart from this, the procedure of taking roots, as indeed a great part of the rules of calculation, are determined by the 'arithmetically unessential' circumstance that numbers are here represented in power series. Cf. what was said in Section II, p. 50.

[21] Russell makes the apposite remark: "From the habit of being influenced by spatial imagination people have supposed that series must have limits in cases where it seems odd if they do not. Thus perceiving that there was no rational limit to the ratios whose square is less than 2, they allowed themselves to 'postulate' an irrational limit, which was to fill the Dedekind gap. Dedekind, in the above-mentioned work, set up the axiom that the gap must always be filled, i.e. that every section must have a boundary. It is for this reason that series where his axiom is verified are called 'Dedekindian'. But there are an infinite number of series for which it is not verified". "The method of 'postulating' what we want has many advantages; they are the same as the advantages of theft over honest toil. Let us leave them to others and proceed with out honest toil." (*Introduction to Mathematical Philosophy*, p. 71.)

[22] R. Dedekind, *Stetigkeit und irrationale Zahlen*, 1872; 5th edition, Brunswick 1927.

[23] This corresponds to Cantor's definition of irrational number by fundamental series.

[24] *Introduction to Mathematical Philosophy*. p. 72.

[25] Cf. note 30 below.

[26] The arbitrariness in the process of narrowing the intervals in constructing irrational numbers forms the basis on which Brouwer builds up his theory of the continuum as a 'medium of free becoming'.

[27] This definition of the equality of two irrational numbers essentially agrees with that given by Weyl, 'Grundlagenkrise', l.c., p. 72, following Brouwer.

[28] Cf. below p. 108f.

[29] The demand that sets be constructible, which Brouwer above all has most emphatically voiced, has doubtless initiated an important advance. Brouwer himself reaches a determination of real number very similar to that given here, as soon as we leave aside the time interpretations that accompany his account. These must absolutely be eliminated, if we are to gain an accurate grasp of the specific meaning of mathematical aspects. We must be absolutely clear that an 'irrational number' is not at all something that 'becomes': it is an abbreviated way of denoting (an incomplete symbol for) a mathematical relation, and the dissolution of the attendant problems lies in a simple description of that relation without attempts at substantivizing definitions.

[30] In Russell's system, these difficulties arise because on the basis of his ramified theory of types he has to assume irrational numbers of different levels, while in the sense of analysis every irrational number must be representable as a sequence of rational numbers (as an infinite continued fraction, infinite decimal fraction). In this way Russell was

forced to introduce his so-called axiom of reducibility, which says that every propositional of a certain level has the same denotation as a propositional function of the first level (that is, they will be true, or false, together for the same arguments). However, quite apart from other theoretical defects in the axiom of reducibility, this completely paralyses the effect of the ramified theory of types on analysis, as has been particularly emphasized by Ramsay (l.c., p. 359). (The effect on epistemological antinomies, however, remains unimpaired, see below p. 161ff.) Since, in fact, *Principia Mathematica* builds up analysis on the basis of the axiom of reducibility, it follows that the ramified theory of types is out of place in analysis. Nevertheless, as he admits in the preface to the second edition of *Principia Mathematica*, Russell has not succeeded in erecting the theory of irrational numbers without the ramified theory of types and the attendant principle of reducibility. A similar attempt by Ramsey (in the work cited above) has remained equally unsuccessful. (I gather from conversation with Waismann that Ramsey himself admits this.) However, these difficulties reside not in the matter itself, but are caused by Russell's class theory of numbers.

[31] That we can determine such a sequence $z_1, z_2, \ldots, z_n, \ldots$ is implied by the hypothesis that $f_1, f_2, \ldots, f_n, \ldots$ is a monotonic increasing sequence of sequences, for this simply means that to any two sequences f_i and f_{i+1} we can assign a rational number greater than every element of f_i but not greater than every element of f_{i+1}.

[32] Cf. Weyl, 'Der circulus vitiosus . . .', l.c., p. 87ff.

[33] Such algebraic equations have no roots that are fractional numbers. However, the irrational numbers that are roots of such equations are called algebraic integers. This concept plays a fundamental part in higher number theory, in the theory of number fields and 'ideals'. Since Hilbert repeatedly cites Kummer's theory of 'ideals' as a paradigm for the cognitive value of introducing ideal concepts (cf. above p. 51ff.), let us state emphatically that we can dissolve the symbolism there used in exactly the same way as we are doing here. Only, in that case the mathematical states of affairs are even more complicated and the simplifying symbolism accordingly even less dispensable.

[34] This restriction too is inessential and is introduced merely to simplify the account; the same holds of the use of the decimal system.

[35] If, for example, the root in question were $3 + \sqrt{5}$, then $p + q = 5.23606 \ldots$ and the two terms $p = 5$ and $q = 0.23606\ldots$

[36] The symbolism of negative numbers could likewise be eliminated without special difficulties.

[37] It is not possible to assign to every arbitrary sequence of natural numbers an algebraic equation in such a way that the relation just described will hold. We therefore distinguish within the domain of irrational numbers those that are roots of algebraic equations and those that are not. The former are called algebraic numbers, the latter transcendental numbers. Concerning these last, cf. Section V, note 13.

[38] We will, however, mention a passage of Felix Klein, in which he rejects the reduction of the irrational to spatial intuition. "That which is given only approximately in intuition or experiment, we formulate in an exact manner, since otherwise there is nothing that we can do with it. This already indicates my position as regards the theory of the irrational. The impetus for forming irrational numbers certainly lies in the seeming continuity of spatial intuition. Since, however, I do not attribute any precision to spatial intuition, I can not allow myself to infer from it that the irrational exists. Rather, I

regard the theory of the irrational as something that must be established or fenced in by purely arithmetic methods. By means of the axioms we can then import these ideas into geometry, in order that there too we may attain the sharpness of distinction that is a prerequisite for mathematical treatment." ('Zur Nicht-Euklidischen Geometrie', *Math. Ann.* **37** (1890), 544–572, p. 572).

[39] Cf. O. Hölder, 'Die mathematische Methode', l.c. p. 199ff.; O. Becker, 'Mathematische Existenz', l.c., p. 476ff.

V

SET THEORY[1]

Our results to date give us the tools for analysing the main concepts of set theory, the mathematical theory of the infinitely large. What is important for this task is above all to distinguish between individual and specific universality, to eliminate the concept of a set in defining natural numbers, to grasp the connection between cardinal and ordinal number, to acknowledge the result of analysing the principle of complete induction and to dissolve the symbolism of irrational numbers.

In particular, we have learnt to see that one may indeed speak sensibly of an infinite set of existing things provided this term is intended to link together things that figure in the same counting process, but that this fact does not cover the meaning belonging to this word where we speak of the 'set of all natural numbers'. For the 'number of elements', which is essential for comparing sets, is given in finite sets by the ordinal number of the last element, while the 'set of all natural numbers' has no last element. As regards this 'set' and indeed any infinite set, we must guard against forming false pictures; the source of all knowledge about infinite sets is the constitutive law of formation.

Once this difference has become clear in principle, we are no longer tempted to operate with infinite totalities as such; we know that statements about infinite sets concern inferences from the law of formation and these alone.[2] We have already observed that statements about a 'general term' are the appropriate form in which to cast the propositions in this area; what can be observed in the general term holds on the basis of the law of formation, everything else remains undetermined (arbitrary). We shall recognize in the sequel that some of the most important statements in set theory, where we seem to be making judgments about the transfinite, become intelligible in their genuine (finite) sense when the 'infinite' is replaced by a range of indeterminacy.

After these introductory remarks let us consider the problems themselves and append our criticism to a brief account of the main theses of set theory.

The set theory created by Cantor aims at providing a mathematical theory of the infinitely large and undertakes to give exact determinations in that domain of the relations 'greater than', 'equal to', 'smaller than' on the one hand, and of relations of 'next bigger' (immediate successor) on the other.[3]

The determination of the former relations (greater than, equal to, smaller than) proceeds with the help of the concepts of a subset and of equivalence. First let us repeat our earlier definition of a subset (p. 83): a set M' is called a subset of a set M, if every element of M' is also an element of M. If M' is identical with M, then M' is called an improper subset of M. Otherwise it is called a proper subset of M.

The definition of equivalence is as follows: a set M is said to be equivalent to a set N if the elements of M can be put in reversible one-one correspondence with the elements of N, that is, in such a way that to every element of M there corresponds one and only one element of N, and conversely.

Given any two sets M and N, the following four cases appear to be conceivable from the start: (1) M is equivalent to a subset of N and N is equivalent to a subset of M. (2) M is equivalent to a subset of N and N is not equivalent to any subset of M. (3) M is not equivalent to any subset of N and N is equivalent to a subset of M. (4) M is not equivalent to any subset of N and N is not equivalent to any subset of M.

In case 1, the so-called theorem of equivalence in set theory states that the sets M and N are themselves equivalent.[4] Of equivalents sets we also say that they are of the same power or that they have the same cardinal number.

In case 2, N is said to be of higher power than M; or alternatively, N is said to have a greater cardinal number than M.

In case 3 the converse holds: M is a set of higher power than N (M has the greater cardinal number).

Finally, case 4, according to set theory, is never to occur. In proving this assertion, set theory rests heavily on the so-called theorem of well-ordering, which we shall have to analyse later.

According to this it is a certain fact for two arbitrary sets whether they are of the same power (have the same cardinal number) or which of the two is of higher power (has the bigger cardinal number).

The practical cognitive import of these definitions accrues if such one-one correlations between sets and their proper subsets are actually performed or if the corresponding correlating principles are indicated. A simple example is the one-one correlation between natural numbers and arbitrary multiples of natural numbers. It can be presented for instance in the following way:

$$\begin{array}{l} 1, 2, 3, 4, \ldots i, \ldots \\ 2, 4, 6, 8, \ldots 2i, \ldots \\ \ldots\ldots\ldots\ldots\ldots\ldots \\ n, 2n, 3n, 4n, \ldots ni, \ldots \end{array}$$

To every natural number we correlate a certain multiple, the same in each instance. Here it turns out (or rather, it seems to turn out) that a one-one correlation between a set and one of its proper subsets can actually be performed. For example, all even numbers are natural numbers but not all natural numbers are even. With finite sets such a representation of a set on a proper subset is excluded. Accordingly, R. Dedekind[5] has raised the equivalence of a set with one of its proper subsets to the status of a defining principle for a set being infinite. The definition runs as follows: "A set M is called infinite (transfinite), if there is a proper subset of M that is equivalent to M. If there is none, then M is called finite".[6]

The sets that are equivalent to the set of natural numbers are called denumerable sets.

Simple considerations show that the set of rational numbers can likewise be put in one-one correlation with the set of natural numbers.

The question now is whether it is possible at all to exhibit sets of a higher power than the set of natural numbers? This is evidently a decisive question for set theory, for if the answer were no, then there would be no multiplicity of infinite cardinal numbers.[7] The proof that there are such numbers is conducted by Cantor as follows:

Let there be given denumerably many decimal fractions[8]

$$\begin{array}{l} a_1 = 0.\ a_{11}\ a_{12}\ a_{13} \ldots a_{1m} \ldots \\ a_2 = 0.\ a_{21}\ a_{22}\ a_{23} \ldots a_{2m} \ldots \\ a_2 = 0.\ a_{31}\ a_{32}\ a_{33} \ldots a_{2m} \ldots \\ \quad\vdots \qquad\qquad\qquad \vdots \\ a_n = 0.\ a_{n1}\ a_{n2}\ a_{n3} \ldots a_{nm} \ldots \\ \quad\vdots \qquad\qquad\qquad \vdots \end{array}$$

Now form a decimal fraction $b = 0.\ b_1\ b_2\ b_3 \ldots, b_n \ldots$ in such a way that $b_n \gtrless a_{nn}$, then b is different from any a_i, differing from it in at least one place, so that it cannot be contained in the above series.

This famous proof procedure is called the diagonal procedure, because the new decimal fraction is formed by replacing every numeral in the diagonal by a different numeral. Cantor considers that this proof establishes the theorem that the set of all decimal fractions is of higher power than the set of natural numbers, since the proof shows that in whatever way we put all natural numbers into one-one correlation with infinite decimal fractions, we can always exhibit a decimal fraction not encompassed by this correlation, while on the other hand the set of all integers can be put into one-one correlation with a subset of all decimal fractions.

Let us interrupt our exposition at this point in order to proceed to a critique of the set-theoretical theses so far described. We begin with the analysis of those correlations that are regarded by set theory as one-one representations of the set of natural numbers on subsets of that set.

The existence of such a one-one correspondence though seemingly indisputable is nevertheless somewhat paradoxical; for on this basis the whole would no longer be greater than any of its parts. From this Leibniz[9] had already inferred that "the number or set of all numbers contains a contradiction, if one takes it as a totality". For Bolzano too, whose *Paradoxien des Unendlichen*[10] must be regarded as the first and only precursor of Cantor's ideas, it is precisely in this correlation (especially emphasized in §20) that a paradox of the infinite lies. However, set theory interprets this state of affairs as meaning that the semblance of paradox arises only if we insist on extending the principle of the whole being greater than any of its parts beyond the finite domain where it does hold; accordingly, set theory uses precisely the division created by the validity or non-validity of that axiom, in order sharply to separate by definition the domains of the finite and the infinite.[11]

Do the correlations indicated actually show that we can give a reversible one-one correlation of a set with one of its proper subsets? It is this assertion that we emphatically deny.

In order to see clearly here, we must most meticulously distinguish between the existing state of affairs, which must be rendered in plain terms, and any interpretations of it. Then we recognize that what shows itself for example in a one-one correlation between the natural numbers and even numbers is merely that to any arbitrary natural number n we can reversibly and unambiguously assign an even number. By contrast, it does by no means show itself that 'therefore' the 'set of all natural numbers' can be reversibly and unambiguously be represented on the 'set of all even numbers'. If of course all that we mean is the state of affairs described in the previous sentence, then everything is in order; but because of the ambiguity of the concept of a set, arising from a fusion of individual and specific universality, an interpretation going beyond this state of affairs readily suggests itself, and that leads to fatal consequences.

We here have a train of thought typical of set-theoretical speculation, and we shall presently describe its general structure. First, one indicates mathematical laws concerning one-one correlation between numbers or the ordering of numbers; next, one interprets these laws with the help of transfinite totalities, after fusing specific and individual universality. By this procedure the transfinite totalities seem to acquire a precise mathematical sense, since one imagines that one has ascertained law-like connections existing 'between them'; indeed, it seems possible to define these totalities as the very 'bearers' of these law-like characteristics (after the pattern of Dedekind's definition of the infinite). Up to this point we can raise 'only' extra-mathematical objections against this train of thought; for even though it is wrong to fuse specific and individual universality, in the mathematical procedure everything must be correct as long as we do not go beyond the foundations being mathematically interpreted, namely the mathematical relations themselves. Therefore, up to this point every assertion about the infinite can be 'translated' into a correct mode of expression; the infinite can be regarded as a mere 'façon de parler'.[12] Yet the dangers of this interpretation have their effects in mathematics itself as well. For a transfinite interpretation of mathematical procedures can introduce incorrect views about the internal mathematical character of these procedures, and in the sequel there will be 'extensions' of such procedures that take their support from these misinterpretations, without finding any justification in the cognitive facts themselves. The most striking example of a train of thought on the pattern just described is to be found in a fundamental aspect of set theory, namely Cantor's theory of the power sets of infinite sets in connection with his interpretation of the diagonal procedure.

In principle it will turn out that we can always find a mathematical sense for statements of set theory about denumerably infinite sets, but never for statements about non-denumerably infinite sets, unless the non-denumerable infinite can be eliminated from these statements (see below, p. 138).

We now return to our example of the one-one correlation between the natural numbers and certain multiples thereof. That such a principle of correlation can be specified is due to the fact that on either side there is a first, second, *n*th element, but no last element. The one-one correlation is effected precisely by pairing the respective *n*th elements. In analysing the concept of natural number we have already discussed this connection between correlation and order, and need not repeat it here. The result was that the fact of a one-one correlation being realisable is independent of the way in which the elements are arranged, but that we do presuppose some arrangement or other.

This can likewise be seen in the representation of the rational numbers or the algebraic numbers on the natural numbers[13], since this representation is effected by ordering the rational numbers p/q (with p and q relatively prime natural numbers) according to the sum of numerator and denominator $p + q$, or by ordering algebraic numbers in terms of the 'height'[14] of the equations they satisfy. It is this order that makes them 'denumerable'. Indeed, as soon as one frees oneself from the sham notion of an infinite totality being 'present' independently of a law of formation, it becomes evident that arbitrarily many elements can be reached, in the temporal sense, only by means of a 'generating order', that is, by means of a principle that allows us to determine an $(n + 1)$th element from the previous n elements. In contrast, classical set theory, which regards an infinite set as given by a totality of elements, just as a finite set is, imagined that it could completely separate the principles of one-one correlation between the elements of sets (the doctrine of the order of magnitude of transfinite cardinal numbers) from the principles of the arrangement of the elements, and then afterwards link the two kinds of consideration in the theory of ordering and well-ordering. In fact, however, in meaningful and mathematically correct set-theoretical statements about infinite sets the arrangement of the elements is not denied, but merely left open as regards the special manner of their arrangement. What we said in Section III about finite cardinal numbers here applies with the further restriction that finitely many things can indeed be present without being ordered by counting, but an analogous 'being present' of the infinite is excluded.

This state of affairs becomes clearer still on closer analysis of the definition of equivalence, which operates with the concept of 'one-one correlation' or with our being able to effect such a correlation. The potential aspect contained in this terminology evidently expresses the fact that there is here a question not of anything depending on the current stock of knowledge, but of an objective criterion. For example, the 'totality of even numbers' and the 'totality of natural numbers' can be put into reversible one-one correlation only by determining for any arbitrary even number an immediate even successor, so that there is a first, second, . . . nth even number. The one-one correlation between the 'set of natural numbers' and the 'set of rational numbers' is to be understood in an analogous manner. It is thus not as though there were 'first' a 'given' and unordered totality of ordered pairs of natural numbers[15], that 'next' a law could be discovered by means of which one could determine a first pair and for every pair an immediate successor pair, and, finally and in the third place a one-one correlation with the natural numbers were effected on the basis of this law. On the contrary, the position is as follows: the concept of an infinite totality of number pairs has meaning only in connection with such a law of formation, the specification of which is simply the reversible one-one correlation with the natural numbers, just because the law determines a first, second, . . . nth number pair.

The false assumption of the actual infinite as a totality of discrete elements leads to important consequences in the interpretation of the diagonal procedure, to whose analysis we now proceed.

We have recognized that the finite sense of propositions about one-one correlations between denumerable sets consists in the stating of a principle that effects, for any arbitrary natural number n, the correlation of the nth elements of the two sets.

In the diagonal procedure, on the other hand, what is proved is that a one-one correlation of a certain kind (presently to be described) cannot be effected. Let several numbers be given (say the numbers 0 to 9), and from these let n variations of the nth class be formed (with n a natural number > 1); then there are evidently further variations of the nth class, since their total number 10^n is greater than n. In order to obtain such a variation we can proceed in the following manner: put the n existing variations into some order and form that variation which is made up of the first term of the first variation, the second term of the second variation, . . . the nth term of the nth variation. Then we form a further variation of n terms from the given numbers (0 to 9) in which the first, second, . . . nth terms are different respectively from the corresponding terms of the variation just previously described. The variation so constructed is certainly different from any of the initial n variations. This serves to mark the finite principle underlying the diagonal procedure.

In the previous section we have shown, in connection with the analysis of irrational numbers, that an infinite decimal fraction is merely a symbol for a denumerable sequence of natural numbers that satisfies certain mathematical conditions. If we leave aside the symbolism of decimal fractions, which is in principle unessential, the diagonal procedure shows how any arbitrary denumerable sequence of denumerable sequences of natural numbers can provide definitions for further denumerable sequences of natural numbers not contained in the initial sequence of sequences.[16] This occurs in such a manner that the aspect in which the new sequence differs from any of the initially given sequence enters into its definition; a result achieved by putting every place of the new sequence into one-one correspondence with a certain initial sequence and within the latter to a certain place, and defining the place in the new sequence as different from the place to which it has been correlated.

Given what we have observed so emphatically above, we should hardly need to stress again that the concept of 'denumerable sequence of natural numbers' does not denote an infinite totality of elements existing as such, but the domain of a certain law which is satisfied by any of these natural numbers. If we keep this constantly in mind, we can confidently use the usual terminology of set theory, in order not to make the account more difficult. In the same way, once we have clarified the precise meaning of 'infinite decimal fractions' we can safely operate with this symbolism, in order to connect it directly with our critical account of the diagonal procedure. The question whether the diagonal procedure says something about the 'set of all infinite decimal fractions', a query that has become topical through the diagonal procedure and is decisive for set theory, can now be formulated thus: can such a law of succession of infinite decimal fractions indicate that no infinite decimal fraction is conceivable that is not unambiguously determined by this law?

If such a law existed, we could, on the basis of the diagonal procedure, assert that it is impossible to produce a one-one correlation between all the decimal fractions generated by this law of formation and the natural numbers; and this, together with the equivalence of the set of natural numbers with a subset of the set of decimal fractions, would yield the higher power of the set of decimal fractions as compared with that of the set of natural numbers. However, the diagonal procedure does not constitute such an all-embracing generating principle, and we shall even prove that none such can be found.

Therefore the diagonal procedure by no means proves that a set of higher power than that of the set of natural numbers 'exists' as a mathematical object. The opposite view currently prevalent in set theory can be traced back essentially to the fact that the definition of decimal fractions through the specification of the 'properties' they have in common was taken to be the defining principle of a set. As against this we stress again that the definition of decimal fractions in no way effects their 'collection into a set'; not even in the sense in which set theory speaks of the set of all natural numbers. For the series of natural numbers is defined by a 'generating principle', whereas the totality of decimal fractions is not.[17]

Here, on the threshold of the doctrine of the non-denumerable infinite, let us once more dwell in full awareness on the way in which this doctrine is connected with the misinterpretation of the infinity of the number series.

We have observed that the natural numbers are logical abstractions of the counting process, but that beyond this abstraction the concept of the number series contains an 'idealisation', which consists in the hypothesis that there is no fixed upper bound so that by 'number series' we must understand the abstract notion of an infinite counting process. If one always keeps this in mind and therefore avoids the mistake of seing the number series as a closed totality of natural numbers, then likewise we will no longer be tempted to regard a totality of all subsets of the set of all natural numbers as something given that exists as such (or, if one does not do this, to think of these subsets as being generated by cognition, on an interpretation that misses the sense of cognition). As against this, the misinterpretation of the concept of the 'number series' leads to the misinterpretation of the diagonal procedure. That is why a thorough critique of method must start at this point, even though the logical absurdities arising from the view criticised appear only when we operate with domains of the non-denumerable infinite.

The erroneous view that all infinite decimal fractions between 0 and 1 are 'present', and that we need merely subject this 'material' to mathematical treatment, is further strengthened by the seemingly intuitive character of the linear continuum whose points are to correspond to the real numbers between 0 and 1 (which coincide with proper infinite decimal fractions). In Section IV we have already pointed out how deceptive this appeal to intuition is.[18]

We have observed that the finite sense of the diagonal procedure is based on the cognitive fact that the number of 'variations of the kth class of n elements with repetition', namely n^k, is greater than n if $k > 1$. Since we here have powers of n, it seemed a likely notion actually to 'generate' higher powers for sets by this process of potentiation. This attempt is made by Cantor by forming covering sets of denumerable sets. Let us clarifv his train of thought by using once again the example of infinite decimal fractions.

Two proper infinite decimal fractions that do not break off are different from each other if and only if they differ in at least one place. If now each place after the decimal point is varied from 0 to 9 (or, as Cantor puts it, 'is covered by the numbers from 0 to 9'), it seems evident that there is no decimal fraction that is not 'generated' by this procedure; and so a method would be given for determining 'all proper infinite decimal fractions'.

However, is the formation of a 'covering set'[19] in fact a 'procedure'? Can we actually construct the covering set from a denumerable set; that is, can we define it without the help of the previously criticized principle of comprehension? For this to be the case it would be necessary for a covering order to be present, by means of which an $(n + 1)$th covering appears unambiguously determined on the basis of the n previous coverings; since without such a rule the prescription for covering 'to infinity' lacks any sense whatever.

To clarify the relations here at work let us compare this case to the 'set of all natural numbers'. In the latter, repeating the same operation (addition of 1) leads us from an arbitrary initial number to every number greater than it; but in the sham procedure of covering, such a law-like character is lacking. Even to pose the problem of finding a procedure (law of formation) for 'all infinite decimal fractions' is in this formulation inadmissible. For one makes the forbidden prior assumption that independently of the law of formation we have 'present' before us all the decimal fractions that are to be linked by means of that law. We may ask only whether a law of formation (possibly consisting of several partial laws) can be specified such that for any two given decimal fractions it shows how they are connected by that procedure.[20]

This question already leads us into the second main area of pure set theory, namely the theory of well-ordering. Before turning to it, let us point out that Cantor reaches ever higher powers (transfinite cardinal numbers) by iterating the formation of covering sets or sets of all subsets of infinite sets. The proof is obtained with the help of a generalisation of the diagonal procedure.

Thus in the sense of Cantor's theory the set of all real numbers is the power set of the set of all natural numbers, and the set of all unambiguous real functions $f(x)$ is the power set of all real numbers.

Moreover, we mention the fact that Cantor 'calculates' with transfinite cardinal numbers in a similar way as with finite numbers (adding, multiplying and potentiating).[21]

For example, if we denote the cardinal number of the set of all natural numbers by a, the cardinal number of the set of all real numbers (the continuum) by c, and the cardinal number of the set of all real functions by f, we obtain the following equations, (where n is an arbitrary finite number):

$a + n = a$	$a.n = a$	$a^n = a$
$a + a = a$	$a.a = a$	$n^a = c$
$c + a + n = c$	$c.a = c$	$a^a = c$
$c + c = c$	$c.c = c$	$c^n = c$
$f + c = f$	$f.f = f$	$c^c = f$

Translated into geometrically 'intuitive' terms, $c.a = c$ for example states that the set of all points of an arbitrary distance is equivalent to the set of all points of a given segment of a straight line, and $c^n = c$ states that the set of all points of a one-dimensional continuum is equivalent to the set of all points of a continuum of an arbitrary number of dimensions.

Finally let us emphasize even at this early stage that the formation of a power set does indeed represent the ascent to a higher power (cardinal number) in the sense of set theory, but that it is not certain whether that higher power is the next higher power (cardinal number).

However, Cantor reaches the next higher power by means of his theory of well-ordered sets (or ordinal numbers). The problems arising from this double-tracked procedure will have to be considered later.

We now come to a brief account of the principles of well-ordered sets, or the theory of ordinal numbers. Let us start with well-ordered sets.

Since set theory maintains the thesis (previously mentioned and criticized) that one can perform mental operations with unordered infinite sets, it is faced with the problem of stipulating by definition under what conditions an infinite set is to be called ordered. This is achieved by defining order as a connected, asymmetric, and transitive relation between the elements of a set. That is to say: (1) of the two elements a and b one precedes the other (connection); (2) if of two elements, a and b, a precedes b,[22] then b cannot precede a (asymmetry); (3) if of three elements a, b, and c, a precedes b and b precedes c then a precedes c (transitivity).

The elements of a set consisting of at least two elements, and a fortiori of an infinite set, can always be ordered in different ways. For example, the set of whole (positive and negative) numbers can be ordered in the following ways:

$$(\ldots -4, -3, -2, -1, 0, 1, 2, 3, 4, \ldots)$$
$$(0, 1, -1, 2, -2, 3, -2, \ldots)$$
$$(0, 2, -2, 4, -4, 6, -6, \ldots 1, -1, 3, -3, 5, -5, \ldots)$$

The definition of the same order or, as it is called in set theory, the 'similarity' of ordered sets is arrived at as follows:

An ordered set M is said to be similar to an ordered set N if the elements of N can be correlated with those of M in such a way that to every element m of M there corresponds, in reversibly unambiguous fashion, one and only one element n of N, and that in this correlation the order of corresponding elements is preserved (that is, if m, n and m', n' are two pairs of corresponding elements, from the relation 'm precedes m'' holding in M, the relation 'n precedes n'' holding in N always follows, and conversely).

The similarity of sets thus includes their equivalence. Of two similar sets we also say that they belong to the same ordinal type. The ordinal type of the natural numbers in the 'natural' order 1, 2, 3 . . . is denoted by ω, and the ordinal type of the negative numbers in the 'natural' order . . . -3, -2, -1 by $^*\omega$.

A special kind of ordered set in set theory are well-ordered sets, whose concept is fixed by the following definition:

A set is called well-ordered if every subset of M different from the empty set (and therefore M itself) has a first element. For example, sets belonging to the ordinal type ω are well-ordered, whereas sets of ordinal type $^*\omega$ are not well-ordered.

A further example of well-ordering is the arrangement of the set of positive and negative integers in the following manner:

(0, 1, 2, 3, . . . -1, -2, -3 . . .), but the arrangement
(. . . -3, -2, -1, 0, 1, 2, 3, . . .) is not well-ordered.

From the definition of well-ordering the following propositions follow immediately:

(1) In a well-ordered set every element (except the last, if any) has one and only one immediate successor. (2) Every ordered set that is similar to a well-ordered set is itself well-ordered. (3) Every subset of a well-ordered set is itself well-ordered. (The ordinal type of well-ordered sets are called ordinal numbers or order numbers).[23]

The concept of 'segment of a well-ordered set' is now defined as follows: if m is an arbitrary element of a well-ordered set M, then the subset of all the elements of M preceding m is called the segment of M determined by m. It can be proved that a well-ordered set is not similar to any of its segments.

From this we obtain the order of magnitude of ordinal numbers. For the ordinal number of M (and therefore of any set similar to M) is said to be greater than the ordinal number of any segment of M, and conversely the ordinal number of any segment of M smaller than the ordinal number of M itself.

The 'next biggest' ordinal number is defined as follows: if μ is an arbitrary ordinal number and the set of all ordinal numbers that are smaller than μ is ordered according to the magnitude of its elements, so that it begins with 0, 1, 2, . . ., then the set $W(\mu)$ is well-ordered and has the ordinal number $\mu + 1$.

Cantor saw in this theory of his a continuation of the ordinary number series beyond the finite. Besides, we easily recognize that the natural numbers are ordinal numbers in the sense of our definition; they are finite ordinal numbers, distinguished as such from infinite (transfinite) ordinal numbers, and coincide with the finite cardinal numbers. The ascending series or ordinal numbers beginning with 0 is well defined up to any arbitrary finite ordinal number, by the procedure of obtaining the ordinal number $\mu + 1$ from the ordinal number μ, as just described.

How, then, does Cantor go from finite ordinal numbers to the smallest transfinite ordinal number? For this, a new principle is required, which he formulates with the help of the concept of fundamental series. This concept is defined as follows: if M is an ordered set, every subset of type ω contained in it is called an ascending fundamental series, and every subset of type $^*\omega$ a descending fundamental series.

The new generating principle is this: for every fundamental series $\{f_\nu\}$ of increasing ordinal numbers there exists a smallest ordinal number that is bigger than all f_ν. It is called the limit number of the fundamental series.[24]

Thus ω itself is the limit number of the well-ordered set of natural numbers and therefore the smallest transfinite ordinal number. On the basis of the first principle we then proceed to further ordinal numbers $\omega + n$ (where n is a natural number). The ordinal number $\omega.2$ is again the limit number of the fundamental series ω, $\omega + 1$, . . . $\omega + n$, . . . An example of the ordinal number $\omega.2$ is given by the series 1, 3, 5, 7 . . .; 2, 4, 6, 8, . . .

From these examples we can already see how, in general, the ordinal numbers $\omega.m + n$ are determined.

The ordinal numbers $\omega.\omega$ or ω^2 is then the limit number of the series ω, $\omega.2$, $\omega 3$, . . ., $\omega.n$, ...

The natural numbers can be ordered according to the ordinal type ω^2 as follows: in the first instance, the ordering criterion is taken to be the number of prime factors (where equal prime factors are counted according to their multiplicity); in the second instance, the criterion is the magnitude of the numbers to be ordered. We then obtain the following well-ordered schema:

1, 3, 5, 7, 11, . . .; 4, 6, 9, 10, 14, . . .; 8, 12, 18, 27, . . .;
16, 24, 36, 40, ...

By combining the two Cantor principles indicated, we can form ever higher ordinal numbers. Thus ω^ω arises from the fundamental series ω, ω^2, ω^3, . . ., ω^n, ...

We will now describe a procedure, due to G. Hessenberg, for a well-ordering of all natural numbers according to the ordinal type ω^ω. First, arrange the natural numbers according to the number of (equal or unequal) prime factors contained in them; if the number of factors is the same, then secondly, order the factors according to magnitude in such a way that those numbers precede which contain the smallest prime factor; amongst series that now belong together, the third ordering criterion is to be the highest powers at which those smaller prime factors appear, and fourthly and finally order the numbers in those groups according to magnitude. This yields the following arrangement of the natural numbers: 1, 2, 3, 5, 7, 11, . . .; 4, 6, 10, 14, . . .; 9, 15, 21, 33, . . .; . . . 8, 12, 20, 28, . . .; 18, 30, 42, 66, . . .; . . . 27, 45, 63, 99, . . .; . . . 16, 24, 40, 56, . . .; . . .

The next characteristic stage is then the ascent to the so-called 'ϵ-numbers', which is based on the fundamental series ω, ω^ω, ω^{ω^ω}, . . . $\omega^{\omega^{\omega^{\omega^{\cdot^{\cdot^{\cdot}}}}}}$ . . . .

The introduction of the notation is required because the limit number (denoted by ϵ) of the above series can no longer be represented by using the symbolism of addition, multiplication and potentiation. Those ordinal numbers ξ that satisfy the relation $\omega^\xi = \xi$ are called in general epsilon numbers by Cantor. ϵ itself is the smallest epsilon number. The first construction of ordinal numbers leading to epsilon numbers is due to G. H. Hardy.[25]

We return to the account of the theory of well-ordering. It culminates in the two 'main laws of well-ordered sets' (asserting that any two arbitrary well-ordered sets can be compared as to their ordinal numbers and their cardinal numbers respectively) and in the so-called well-ordering theorem.

The first main law runs thus: two well-ordered sets are either similar or one is similar to a segment of the other. Of two unequal ordinal numbers one is therefore always the smaller and the other the greater.

The second main law runs as follows: Well-ordered sets can be compared not only as regards their ordinal numbers but also as regards their cardinal numbers: the cardinal numbers of two well-ordered sets are either equal, or one of them is smaller than the other.

For if M and N are well-ordered sets and the ordinal number of M is smaller than the ordinal number of N, then the cardinal number of M is equal or smaller than the cardinal number of N.

The second main law, which here interests us particularly, follows readily from the first if we keep in mind that on the one hand the concept of similarity contains the concept of equivalence, while on the other hand the 'segment of a well-ordered set', by means of which the ascent to the next higher ordinal number is effected, is a special subset of that well-ordered set.

However, the two main laws do not yet settle the following question which is of fundamental importance for set theory: it is possible to find, for any transfinite cardinal number, at least one ordinal number that 'belongs' to it? In other words: is it possible, for an infinite set of arbitrary power, to find an equivalent well-ordered set? The existence in principle of this possibility is asserted in the well-ordering theorem: every set can be brought into the form of a well-ordered set.

Mathematicians have summoned up admirable ingenuity in order to prove this theorem, and Zermelo has furnished two 'proofs'[26] which operate with the selection of elements and subsets from infinite sets on the basis of the so-called 'principle of choice'.

If now, so the argument continues, the well-ordering theorem is to hold on the one hand and the calculus of powers on the other, then there must be, amongst ordinal numbers, a smallest ordinal number that belongs to a set that is no longer denumerable, and therefore there must also be a smallest non-denumerable cardinal number. Cantor arrives at this ordinal number in the following manner: he imagines a well-ordered set of all finite and denumerable ordinal numbers. This set in turn has the next higher ordinal number, which can no longer be denumerable and therefore represents the smallest ordinal number belonging to the next-higher power. By iteration of this procedure Cantor travels to ever higher powers, more precisely from one power to the next higher power in each case.

The cardinal numbers of infinite well-ordered sets are denoted as aleph numbers ($\aleph$) with running indices: $\aleph_0, \aleph_1, \aleph_2, \ldots, \aleph_i, \ldots, \aleph_\omega, \ldots$; $\aleph_0$ belongs to denumerable sets. The set of all ordinal numbers belonging to an aleph ordered according to their magnitude is called the number class of this aleph. The finite ordinal numbers are ranged into the first number class, the ordinal numbers belonging to $\aleph_0$ into the second number class.

The further question now arises, what is the relation between the ascent to higher powers with the help of the well-ordering calculus and the ascent with the help of the formation of power sets as described and criticized above. The former, by hypothesis, certainly leads to the next-higher cardinal number, but it could be that in the formation of power sets cardinal numbers might be passed over. What stands in the forefront of interest here is understandably enough the first step in the infinite domain, namely the question whether the power set of denumerable sets, which has the same power as the so-called continuum, has the power $\aleph_1$ or some higher power. This famous unsolved (and, as we shall show, unsolvable) problem of set theory carries the name of problem of the continuum.[27] For decades some of the finest mathematicians have bent their efforts towards mastering this problem.

We will now proceed to a critical analysis of the theory outlined. Our leading principle in this will be to sever, in the sharpest manner, the mathematical states of affairs themselves from the interpretations attached to them. As already mentioned, Cantor sees in transfinite ordinal numbers a continuation of the series of natural numbers beyond the finite. This view, as we shall show, leads him to consequences which are logically untenable, however ingenious the mental conceptions involved. We, on the contrary, shall eliminate the impossible notion of operations in the infinite domain and ask for the genuine finite principle that lies at the base of the step-like sequence of ordinal numbers. This is not a matter of a 'finite interpretation' to be put in the place of Cantor's 'transfinite interpretation', but of a plain account of the mathematical relations themselves. Just one further preliminary remark: with mathematicians one often encounters the following view: 'what the nature of natural numbers (or of any other mathematical object) is does not concern us; the only thing that interests us is the relations taken to hold between natural numbers (or the mathematical objects in question)'. In fact, however, the position is that there is no mathematical object 'behind' those relations; for, as we explained in detail in analysing the concept of natural number in Section III, these relations are incompatibility relations and the mathematical object is simply the 'bearer' of these incompatibilities. 'To ascertain the nature of mathematical objects' thus simply means clearly to grasp the specific connection of these incompatibility relations.

If, then, we ask what is the legitimate mathematical content of the theory of well-ordering or of ordinal numbers, we pose the problem of ascertaining the plain and finite sense of whatever mathematical propositions appear in that theory in transfinite guise.

Let us first examine what is meant by the statement that a certain number is contained in a certain well-ordered infinite set, let us say in the set of prime numbers arranged in order of magnitude. Evidently, this simply means that this number falls under the law as whose domain the set appears; in our example, that the number is a prime number.[28] The 'translation into finite terms', here illustrated by an elementary example, holds as to its sense for all the iterations and incapsulations of laws that constitute the essence of well-ordering, as we shall presently show. Once grasp this state of affairs in principle, and the meaningful content of the theory of well-ordering is not difficult to determine.

Let us now go straight to the heart of the enquiry and choose a simple model as our example for the state of affairs to be described. Suppose various things T are to be marked by signs. Let the stock of signs consist of 2, 4, 6, . . . in unknown number, although it is to be certain that they exist in closed form; that is, the existence of a sign for a higher number (say, 12) is a sufficient condition for the existence of the sign for any arbitrary smaller even number (for example 10). Moreover the three signs 1, 3, 5, also exist. Now let us make the following stipulation for the order of denoting: first we use up the available even numbers, in order of magnitude; when they are exhausted, we use the three odd numbers also in order of magnitude. It therefore remains undetermined what is the immediate precursor of 1; for the number 3, however, there is indeed a determined immediate precursor, namely 1, but the immediate precursor of that immediate precursor remains undetermined. Thus every number that occupies an unambiguously determined place in relation to the division of indeterminacy must be distinguished from any other sign by means of an ordering criterion.

We recognize without difficulty that the 'indeterminacies' that arise here are strictly correlated with 'laws', for the indeterminacy is the domain bounded by the laws of formation. Hence all the following formulations could be transformed in such a way that instead of 'indeterminacies' we speak of 'laws' or of 'functions'.

Let us now lay down the following definitions: 1. Two ordering rules 0_1 and 0_2 are to be called 'similar' if 0_2 introduces the same number of indeterminacies as 0_1, and if the number of determined signs following the last indeterminacy is also the same. 2. An ordering rule 0_2 is called 'next higher' to an ordering rule 0_1 if it defines the same number of indeterminacies as 0_1 but the number of determined signs following the last indeterminacy is greater by 1 than for 0_1.

By these stipulations the finite sense of transfinite ordinal numbers appears fixed for any of the $\omega.m + n$.[29] In order to reach the higher ordinal numbers, the principle of successive alignment of indeterminacies as just applied must be combined with the principle of incapsulation of indeterminacies. For example, for the ordinal number ω^2 the number of successively aligned indeterminacies is itself undetermined and to the ordinal number ω^ω there corresponds the incapsulation of four indeterminacies.

Let us lay bare the logical core of the relations prevailing here by examining the last-named ordinal number, using Hessenberg's previously mentioned example of well-orderings of the natural numbers according to that ordinal type. As we have observed, we have here four incapsulated ordering criteria, namely: 1. Number of prime factors, 2. magnitude of prime factors, 3. highest power of prime factors, 4. magnitude of the numbers.

The mutual order (coming before or after) for any two arbitrary natural numbers is thus fixed with the help of these four criteria, and for any given number it is definite which is its immediate successor.

However, this logical relation must not be interpreted, as happens in set theory, as though certain numbers (say, the number 10) were preceded by iterations of infinite sets of numbers. Rather, we must say: if any two numbers are to hand, there is between them an ordering relation (which could for example be immediate succession) arising from the four criteria mentioned. Which numbers are to hand does, however, remain undetermined; indeed, even for any of the sequences of elements, sequences of sequences, and sequences of sequences of sequences, determined by our four criteria, it remains undetermined how many (or even if any) representatives are to hand.

For the sake of rendering the case 'intuitive' we may regard this 'being to hand' once more in such a way that the numbers are conceived as being represented by numerals; but this must not conceal the insight that the law-like connection in question is non-intuitive in character.[30]

In contrast, Cantor's principle of generation of ever higher ordinal numbers, by which for every fundamental series of increasing ordinal numbers there is an ordinal number greater than all those in the fundamental series, must be rejected in this form, because it operates with infinite multiplicities. The same holds for the above-mentioned[31] corresponding formulation by Cantor which employs well-ordered sets.

The sense of that formulation does, indeed, reach as far as the successive alignment and incapsulation of laws, and every new limit number represents a new step within this process. However, it fails at the point which is decisive for the transfinite theory of sets, namely where it is necessary to give a logical formulation of the ascent to the higher aleph numbers and so to transfinite number classes that go beyond the denumerable.

In analysing the diagonal procedure above, we have shown that this proof by no means guarantees the 'existence' of higher transfinite powers, observing further that the ascent to higher powers by 'forming' power sets can be given mathematical sense only insofar as we succeed in fixing an unambiguous covering order in such a way that on the basis of n coverings we can determine an $(n + 1)$th covering.

Thus we could operate meaningfully with the concept of a set of power higher than denumerable only if it could be shown that the progression in the series of ordinal numbers themselves (that is, the iteration of successive alignments and incapsulations of laws) leads beyond the denumerable (that is, to sets that cannot be one-one correlated with the set of natural numbers while they possess proper subsets for which this can be done).

In brief: at no point in the theory of transfinite ordinal numbers is it permissible to presuppose an independently existing sequence of transfinite cardinal numbers. The double-track in the transfinite process falls away. Yet it is precisely on this presupposition that Cantor's foundation of the sequence of transfinite number classes rests: his theory is therefore invalid.

This shows that the only path that Cantor indicates for reaching higher aleph numbers cannot be travelled. We must, however, still prove that such an ascent is in principle impossible.

Let us make clear to ourselves what it would mean if there were some procedure that made such an ascent to a higher power possible. There would evidently have to be statements of pure mathematics that were valid only in the domain of these higher powers. For mathematical objects are distinguished only by the differences of the logical relations as whose 'bearers' they are defined[32]; since the ordinal numbers that would belong to higher number classes than the second are to be 'new' numbers (that is, different from any finite and denumerable ordinal numbers), certain consistent systems of mathematical statements would have to exist that held for these 'new' numbers but not for any number of the first or second number class. Accordingly, the proof that such systems of mathematical statements are unthinkable would mean the elimination of cardinal numbers going beyond $\aleph_0$.

This proof has in fact been given, following L. Löwenheim,[33] by Th. Skolem,[34] and it is characteristic for the strong immunisation of set theoretical enquiry against antinomies that this proof has not provoked much greater disquiet than has actually been the case.

The Löwenheim–Skolem theorem runs thus: given an infinite series of counting statements u_1, u_2, . . . numbered with the integers; if the requirement that all these statements hold together is consistent, then they can be simultaneously satisfied within the infinite series of the positive numbers 1, 2, 3, . . . with suitable choice of the symbols for classes and relations.

By a 'counting statement' we here understand a statement that is formed from the basic mathematical objects (which we may here assume to be the natural numbers), by means of logical relations such that the concepts of 'all' and 'there exists' occurring in them refer exclusively to the basic objects themselves, but not to classes (or properties) and relations of basic objects. That in spite of these restrictions counting statements encompass the whole domain of consistent mathematical statements clearly appears from the analyses of this article which are directed against the extended functional calculus.[35]

In his proof Skolem adopts the symbolic logic of Schröder and in his formulation he uses the concept of the denumerable infinite in the manner customary amongst mathematicians, which suggests an infinite totality of independent objects. However, the theorem and its proof are not thereby deprived of their validity and they can easily be 'finitised'.

Thus, as Löwenheim and Skolem have shown, arbitrary systems of counting statements are satisfied in the denumerable domain; this being so, any attempt to build up formally independent systems of the non-denumerable infinite (in the sense of Cantor) with the help of the denumerable infinite appears to be condemned to failure from the very start.

What has just been said holds especially for the various axiom systems of set theory. For such systems must consist of counting statements; therefore what they actually describe (against the intention on which they rest and the terminology that Cantor uses) are exclusively relations within the framework of the denumerable.[36] This at once removes the ground from underneath any non-denumerable infinite domains. For if at first blush it might seem possible that such domains exist, in spite of being demonstrably unaccountable for in terms of axiom systems, the observations of Sections I and II above have shown that this possibility drops away.

If however, contrary to the results of our analyses, one perceives in the diagonal procedure a proof for the existence of non-denumerable transfinite domains, the question arises whether the Löwenheim-Skolem antinomy can be squared with the diagonal procedure.

Fraenkel, who has investigated this question,[37] comes to the conclusion that in the proof of Skolem's theorem no allowance is made for non-predicative procedures, and without these one could indeed not go beyond the domain of the denumerable. If, however, we do not allow ourselves to be blinded by linguistic illusions and instead we consider the states of affairs themselves that are intended with the help of the linguistic symbolism, then we recognize that non-predicative 'procedures' or non-predicative concept formations are meaningless.

By a 'non-predicative concept formation' one understands "quite generally the formation of two concepts in such a way that the definition of either necessarily involves the other".[38] If we eliminate the domain of 'concepts' interposed between thinking and language on one side and the objects thought about or named on the other, we readily see what that account amounts to: the object intended by a sign (pseudo-sign[39]) s_1 is to be completely or partially determined by our indicating the object that is intended by a sign s_2, and this latter object in turn is to be determined by our indicating the former; which is evidently circular. Yet circular 'determinations', insofar as they are circular, determine nothing at all.

In the final section we shall have occasion to consider such non-predicative concept formations further; here we remark in advance, that the illusion of our being able to enlist the help of non-predicative concept formations in order to reach new realms of mathematical objects is intimately connected with the assumption of the principle of comprehension. This assumption arises from the view that a totality of objects determined by a 'property' could contain elements that are definable only with the help of this totality.

Just as it is unthinkable that an element of a totality could be defined only with the help of this totality, so it is unthinkable that a term within a mathematical construction (or, as we might alternatively say, within a mathematical process) could be defined only with the help of this construction (or mathematical process). The seeming difference between these two states of affairs is due to the inadmissible importation into mathematics of the concept of time.

This remark is directed above all against O. Becker's notion of self-transcending constructions, that is supposed to enable us to advance constructively to any arbitrary place in the second number class. In this context he imagines the concept of the second number class as the "concept of a general law of a sequence", as an "empty schema for a possible functional law"[40] and, following a procedure for constructing ordinal numbers given by O. Veblen,[41] thinks that in principle he can in this way obtain the systematic construction of all ordinal numbers of the second class. He determines the constructional type of self-transcending constructions as follows: "What is characteristic of this new type consists in its being in principle unclosed: a determinate constructional principle, however widely conceived, will never lead to the goal, but it is only in the course of constructional activity itself that new instructions for continuing the procedure emerge."[42] "... the correct carrying out of the diagonal procedure, too, is possible only in a self-transcending construction. For while the classical procedure does indeed supply for every given denumerated subset of the continuum a new element of the continuum not contained in that subset, but the subset increased by that element is evidently still denumerable, so that constructively viewed this process would lead to our goal only by means of non-denumerably frequent repetition. Yet such 'non-denumerably frequent' repetition evidently presupposes the direct introduction of an 'absolutely non-denumerable' infinite."[43]

First, as regards the concept of an 'empty schema for a possible functional law', with the best of will I can attach no other meaning to it than that one assumes the concept of the functional law somehow to supply a totality of all functional laws, which would restore the principle of comprehension that we have criticized in the present work (as indeed has O. Becker himself). Even on this assumption, Becker's view does of course differ from classical set theory in that he does not isolate the existence of this empty schema but considers it only in correlation with the several instances of 'filling in' that are to be achieved by construction. Therefore one might perhaps interpret his thesis as requiring the empty schema to be viewed as a regulative principle, as an 'infinite task' in the sense of the Marburg school.

However, even this weakened form of the principle of comprehension does not escape the objections that we have raised against it. For a start this means the disappearance of the schema to be filled in by construction. With the disappearance of the 'regulative principle', the goal that is in principle unattainable but is nevertheless regarded as a guideline, the aspect of being unclosed which is Becker's criterion for self-transcending constructions, likewise drops away.

Thus no construction can ever lead beyond the domain determined by the principle underlying it. To start from the example cited by Becker: in the diagonal procedure, incapsulation of formation laws for sequences of numbers, sequences of sequences of numbers and so on, will lead to more and more new 'mathematical objects', but we must at each stage remain within the framework of the most general formation law according to which the progression runs. The progression is determined as an unfolding of this and of no other law. When we operate with the expression 'and so on', as we do in describing infinite constructions, the stress must therefore lie on the word 'so'; and that is determined by a law. As to the time aspect that plays a certain part in Becker's account, we have already observed that it must be eliminated if one is to make mathematical statements.

If, then, we must renounce the construction of a multiplicity of transfinite powers, perhaps there remains a final way of 'saving' them, namely by directly introducing them; but this possibility, too, drops away if we reflect that such an 'introduction' of a new sign (pseudo-sign) means nothing at all, unless there exists an object that it denotes, which in our case would have to be a formal object, since we are dealing with mathematical objects. However, the Löwenheim–Skolem theorem shows that this object cannot lie outside the domain of the denumerable; and we have shown in Section IV that 'intuition' of the continuum allows of no mathematical application in the sense required.[44]

Finally, as regards the introduction of higher powers as 'ideal elements', we can refer the reader to our general analysis of ideal elements in Section II.

We can thus not ascend to higher powers than $\aleph_0$. In particular it follows from this that no sense can be attached to the concept of the set of all decimal fractions, of the 'number continuum'. In this way the continuum problem[45] likewise vanishes. Set theory cannot lead us beyond the denumerable; statements that fall into this domain, if correctly formulated, relate not to infinite multiplicities but to formation laws, that is, to arbitrary elements of number sequences.

As regards the validity of the well-ordering theorem in the denumerable domain, let us point once more to an observation often repeated above: it is not the case that a denumerable set could be given independently of a formation law that fixes the well-ordering, so that we should 'after the event' have to face the task of effecting a well-ordering, but such a formation law is from the very start involved in the concept of a denumerable set. The conception of infinite sets that are not well-ordered and *a fortiori* not ordered at all, arises only if the arbitrariness of order, presupposed in the calculus of powers, is wrongly construed as independence of any order whatsoever.[46]

This misconception, whose far-reaching consequences we have observed in Section III, involves a train of thought of the following kind: if we consider the set of those natural numbers that satisfy a certain condition (let us say, a given diophantine equation), it seems at first as if the set were to hand 'as such' in unordered form, acquiring an order only in terms of the formation law (in our case, the formula solving the diophantine equation). However, this interpretation is mistaken. What in fact exists independently of the formation law is the cognitive fact that any arbitrary such number does satisfy the given condition (the diophantine equation); but this gathering together, which seems to be involved in the concept of a 'set of all numbers having a certain property' consists simply in the stipulation of a formation law that determines the order of position for every single one of these numbers.

In the domain of 'finite sets of real things' the matter stands no differently in principle, as we have shown above. The individual things exist independently of their being counted or not, but their collection into a set is simply their being counted. Of course there is here the additional idea that heightens the confusion, namely of a 'total intuition', a totality in the spatial sense (a filling of space).

If in retrospect we now ask the question why the formation of a sequence of transfinite powers, which we have proved to be absurd, nevertheless possesses so high a degree of plausibility that even today some of the greatest mathematical thinkers regard it as established fact, we are led back to the general remarks put forward in the introduction of the present work, concerning the typical erroneous pathways of thinking that make 'extravagant' use of 'concepts' (symbols). We there distinguished the following steps: the first consists in genuine mathematical knowledge. Starting from there, the second step involves an interpretation based on the mathematical symbolism, which in a certain direction goes beyond the factual cognitive content. Thirdly and finally, on the basis of this interpretation the symbolism is used even where the cognitive facts that originally determined the meaningful content of the symbolism are lacking.

In the case of set theory the diagonal procedure corresponds to the first step, by which we show unobjectionably that every arbitrary sequence of sequences of numbers determines sequences of numbers other than those contained in the original sequence.

Next we have the second conceptual move, the interpretation that goes beyond the meaning of the diagonal procedure. For, on the basis of the principle of comprehension, tacitly presupposed in most cases, it is now asserted that the first step establishes the mathematical existence of powers higher than that of the totality of natural numbers.

The third step consists in operating with this 'existence of higher powers' independently of the underlying cognitive fact of the diagonal procedure, in a way that regards this procedure as constructively inexhaustible, and enlists it in essential manner in mathematical constructions (ascent to higher number classes).

As against this, the Löwenheim–Skolem theory yields the following insight: if to begin with we hypothetically assume the higher transfinite powers in the sense of set theory, we can show that nothing consistent can be said about them that does not already hold in the denumerable domain, so that the newly introduced concept is redundant, insofar as it is not inconsistent.

In the case of redundancy we have meaningful statements and it is merely the interpretation according to which they concern the non-denumerable infinite, that must be rejected. Such interpretations do indeed give rise to the illusion that the domain of the non-denumerable infinite could be legitimized in logico-mathematical fashion, since one seems to be making meaningful statements 'about it'.

As a paradigm for such a train of thought let us cite the set-theoretical theorem that $c^n = c$. As shown above p. 124f., it states in geometrical formulation, that n-dimensional space has no more points than the one-dimensional continuum, for instance a straight line or an arbitrarily small distance. For simplicity, let us confine ourselves to the assertion $c^2 = c$. Then the proof runs as follows: represent a one-dimensional continuum by the totality of finite and infinite decimal fractions between 0 and 1; then c^2 is the cardinal number of the totality of all pairs of decimal fractions between 0 and 1. Let $0.x_1x_2x_3 \ldots$ and $0.y_1y_2y_3 \ldots$ be any two such decimal fractions, then they unambiguously determine the decimal fraction $0.x_1y_1x_2y_2x_3y_3 \ldots$ which likewise is between 0 and 1. On the other hand any decimal fraction $0.z_1z_2z_3z_4z_5z_6 \ldots$ unambiguously determines two decimal fractions between 0 and 1, namely $x = 0.z_1z_3z_5 \ldots$ and $y = 0.z_2z_4z_6 \ldots$. This proves that an arbitrary pair of decimal fractions between 0 and 1 can be put into reversible one-one correlation with one decimal fraction between 0 and 1. Accordingly the proposition $c^2 = c$ is reduced to the proposition $2a = a$, and generally $c^n = c$ to $n.a = a$.

On the basis of the insights we have gained the nature of this connection is immediately clear. The subject of the proof is the one-one representation of an arbitrary pair (or generally, *n*-tuplet) of proper decimal fractions on one proper decimal fraction; that is, we can establish a relation by which to each *n*-tuplet of proper decimal fractions one and only one proper decimal fraction is assigned, and to each arbitrary proper decimal fraction one and only one *n*-tuplet of them. Expressed in the terminology of set theory, this is a statement about denumerable sets. If, on the basis of the principle of comprehension, we reinterpret our statements about any arbitrary decimal fraction as concerning the totality of all decimal fractions, then what these statements are about (in the sense of the conception currently prevailing in set theory) is the continuum. It is thus obvious that the finitized and meaningful statement 'about the continuum' by no means provides an argument for the legitimacy of that concept.[48]

If, contrary to the view that we have established above, one proceeds from the assumption that Cantor's theses about the transfinite, and particularly about the non-denumerable infinite, contain mathematical knowledge, it seems understandable that mathematicians should strive to free this alleged mathematical knowledge from the absurdities (antinomies) that are well known to arise in its wake, so as to build up a logically impeccable (strict) set theory.

The fittest instrument on offer for these aspirations was the axiomatic method; thus, in 1908, E. Zermelo[49] carried out an axiomatization of set theory, which maintains Cantor's results while eliminating 'paradoxical sets'. In a series of more recent axiom systems, that follow Zermelo more or less closely, the prevailing effort is to eliminate not only the 'classical' paradoxical sets, but also a number of other concepts and methods that seemed suspect, without giving up the essential results of Georg Cantor's set theory.[50]

As regards the capability and limitations of the axiomatic method in general, we refer the reader to our investigation in Section II of the present work. As to set theory in particular, the following additional remarks spring to mind.

In line with our results, which show how corrupt the foundations of the conceptual edifice of set theory are, any system of axioms that seeks to prop up this edifice sets itself an unattainable goal. Nevertheless, the difficult and subtle work that has been done with this aim must not be underrated in its significance for the theory of mathematics as a whole. For by these efforts a number of the most vital problems of mathematical theory first came into the full light of consciousness, and this is usually the most difficult part of the road towards a solution, as with problems in the theory of science in general.

Let us now give the axiom system that A. Fraenkel puts up in his *Einleitung in die Mengenlehre*,[51] and then make some remarks of a general nature about it, without going into a detailed analysis.

Fraenkel's system of axioms contains besides the concepts of formal logic, one single basic relation, denoted by the symbol ϵ.

"Between any two objects m and n of the basic category 'set' given in determinate sequence, the basic relation denoted by the sign ϵ (the first letter of the copula 'ἐστί) is either to hold or not to hold. In the former case we write $m\epsilon n$, or in words: m is an element of n, n contains or possesses the element m, m occurs in n (as an element) and so on; in the latter case, that is, if m is not an element of n, we write $m\bar{\epsilon}n$."[52]

The axioms are preceded by three definitions, which run thus: "Definition 1. If m and n are sets of such a kind that every element of the set m also occurs as an element in n (that from $a\epsilon m$ it always follows that $a\epsilon n$), then m is called a subset of n."

"Definition 2. If m and n are sets and m is a subset of n as well as n of m, then m and n are said to be equal; in signs $m = n$. In all other cases, m is said to be different from n ($m \neq n$)."

"Definition 3. If m and n are sets without common elements, that is, if no element of m occurs also as an element in n, then m and n are called disjoint. If more generally two arbitrary elements of a set M are always disjoint sets, then the elements of M are called disjoint by pairs or simply disjoint."[53]

For the sake of perspicuous presentation (which has no mathematical significance), the axioms are divided into groups.

The first group consists of the following single axiom:

"Axiom I. If a, b, A are sets, a is an element of A and $a = b$ holds, then b is also an element of A (axiom of definiteness)."[54]

The second group, which carries the name of 'extending conditional existence axioms', comprises three axioms, namely: "Axiom II. If a and b are different sets, then there exists a set containing the elements a and b but no element different from these. In view of the foregoing, this set is to be denoted by $\{a, b\}$ and is called the pair of a and b, (axiom of pairing)."[55]

"Axiom III. If M is a set containing at least one element, there is a set containing as its elements the elements of the elements of M, but no other elements (axiom of union)."[56]

"Axiom IV. If m is a set, there is a set that contains as its elements all subsets of m, but no other elements (axiom of the power set)."[57]

The third group is that of the 'restricting conditional existence axioms' and contains two axioms, namely:

"Axiom V. If m is a set and $\mathfrak{E}$ a property that is meaningful (whether applicable or inapplicable) for every individual element of m, there exists a set which contains as its elements all those elements of m that have the property $\mathfrak{E}$, but no other elements. This set is therefore a subset of m, arising from it by 'separation' of the elements having the property $\mathfrak{E}$, and denoted by $m_{\mathfrak{E}}$ (axiom of separation of subsets)."[58]

"Axiom VI. Let M be a set whose elements each contain at least one element and moreover are disjoint by pairs. Then there exists at least one set S, namely a subset of the union $\mathfrak{S}M$, which shares exactly one element with every element of M, but has no other elements. Every such set S is called a selection set of M (axiom of choice)."[59]

The last group finally consists of an absolute existence axiom running as follows:[60]

"Axiom VII. There is at least one set Z having the following two properties: 1. If the null set (that is, a set without elements) exists, then it is an element of Z. 2. If m is any element of Z, then $\{m\}$ (that is, the set containing m and no other element) is also an element of Z (axiom of infinity)."

The question now is, which of these axioms would afford ascent to the non-denumerable infinite and accordingly would be liable to our earlier objections. It turns out, as is well known, that this function belongs to the axiom of the power set, which is why a considerable part of set-theoretical problems centres on that axiom. Since, as we have recognized, the building up of higher transfinite powers is based on the principle of comprehension, we may surmise that essential use is made of this principle in that axiom.

This surmise is confirmed; the axiom of the power set can be derived as a direct consequence of the principle of comprehension. For that a set is 'given', namely unambiguously determined, includes that it is objectively certain for any given object, whether it is or is not a subset of this set. The inference from this definiteness of arbitrary subsets to the definiteness of the totality of subsets of a given set simply amounts to applying the principle of comprehension.

Of course, this holds only if the axiom of the power set is regarded as a pure existence axiom, as with Zermelo. In Fraenkel, by contrast, the position is different, since he wishes to take into account only constructively obtained subsets. For he makes the following remarks about this axiom: "Let us here emphasize once more that, as in Definition 1, so in Axiom IV, the concept of 'subset' has a different and essentially narrower meaning than in Cantor's set theory. In the latter, when forming the power set $\mathfrak{U}m$, we were able to collect an arbitrary totality of elements of m into a subset of m, being certain that this subset would figure amongst the elements of $\mathfrak{U}m$. In the present case we are no longer allowed a 'formation' of a subset of m under such generously free conditions, so that the appearance of the subset in $\mathfrak{U}m$ is by no means guaranteed. On the contrary, a set must first be otherwise given as existing, in order that, according to Definition 1 we might examine whether it is a subset of m; only if in that case the result is positive can we be certain that it figures in $\mathfrak{U}m$."[61]

In this definition of the power set, which satisfies constructivist postulates, it is thus required of each individual subset that it be 'otherwise given'. For denumerable subsets this means that every individual constitutive formation law must be present. 'All' denumerable subsets of a denumerable set, however, cannot be present in this manner, and for that reason the ascent to the 'non-denumerable infinite' in the sense of Cantor's set theory cannot be attained by means of this 'purified' postulate.[62]

Alongside the axiom of the power set the axiom of choice deserves special attention, which it has in fact attracted, indeed more so than the former.

The sense of this proposition is conveyed most strikingly by Weyl's definition as follows: "The principle of choice is the postulate that an existentially determined set can be constructed". As we know, however, existentially to determine a set is to determine it by means of the principle of comprehension. We know further that this postulate cannot be fulfilled if the 'existentially determined set' is non-denumerable. That nevertheless many people regard this postulate as a necessity of thought, as an *a priori* principle (for example Poincaré), is due to the fact that they take the concept of a set always to involve a constructive aspect, even if the concept is to be determined by means of the principle of comprehension.

This becomes clearer still if we take into account that the principle of choice and the well-ordering theorem are logically equivalent; that is, that either of the two assertions follows from the other.[63] For, as we have already emphasized, an 'arbitrary well-ordering' is already involved in the concept of a set. However, these are questions of the psychology of cognition that we do not wish to pursue further.

The following observation, however, is important: since the principle of choice cannot be satisfied beyond the domain of the denumerable, as soon as it is fulfilled all reservations disappear that might justly lead to our being prohibited from using the law of excluded middle in the transfinite domain.[64] By his ϵ-axiom (cf. above p. 64), which is intimately connected with the axiom of choice in set theory, Hilbert secures the applicability of the principle of excluded middle in his axiom system; though he does not thereby attain his goal, namely to secure for himself the tools for a mathematical treatment of the non-denumerable infinite.

We further easily recognize, what is the significance of Hilbert's thesis that against a proposition proved with the help of the axiom of choice a counter-proof can never succeed.[65] For the proof of a proposition *P* by means of the principle of choice states that *P* holds in any proper domain of mathematical objects (not 'generated' merely by the principle of comprehension), for which the other premisses required to prove *P* also hold (that is, all the others except the axiom of choice). If now a counter-proof could be given against *P*, this would mean that one could meaningfully operate with mathematical domains for which the axiom of choice does not hold, which are therefore constituted only by means of the principle of comprehension.

Let us add a few words about the absolute existence axiom, which postulates the existence of certain infinite sets. Zermelo and Fraenkel themselves have clearly distinguished this axiom from the first six, which are called 'axioms of general set theory', thus marking its special position with regard to the others. I should like to assert in addition that such an axiom of absolute existence has no place whatsoever in mathematical axiom systems. For this 'there exists' by no means coincides with the mathematical 'there exists' that we have analysed in Section I of this work, but it intends to assert something about the 'real world' and therefore constitutes an illicit 'shift into another kind'. In rejecting the conception of the denumerable infinite as an actual infinite, the conceptual difficulty that has led to this axiom being put forward will indeed itself disappear.

If, then, in this way we have attained the result that an axiomatic method is equally incapable of giving logical support to the sham edifice of higher transfinite powers so that we are no longer justified in speaking of a set theory existing alongside arithmetic, we should by no means misappreciate the tremendous and as yet unassessable import of those Cantorian discoveries that remain uncontested. What remains includes above all the diagonal procedure with those of its consequences that make no use of the principle of comprehension and the theory of well-ordering insofar as it remains within the denumerable domain. That the transfinite interpretation of these results must be abandoned, says nothing against the extraordinary mathematical fruitfulness of the discoveries themselves. On them rests above all the doctrine of set-theoretical topology; founded by Cantor himself, it has in recent decades grown into a special mathematical discipline, latterly leading to unimagined results particularly by the establishment of the theory of dimensions and the theory of curves. If one eliminates the non-denumerable infinite, the gist of these theories remains uncontested;[66] it is only that certain changes in name may show themselves to be expedient.

Accordingly, we may well assert that the part of G. Cantor's work that remains uncontested is sufficient to secure for him a place of honour amongst the very greatest mathematicians.

In concluding this section, let us once more express our view, already indicated in the introduction, that a recognition, on the part of the leading theoreticians of mathematics, that the non-denumerable infinite is logically untenable 'is in the air'. As regards Brouwer and Weyl, this needs no further confirmation; let us, however, mention a few passages relating to Russell and Hilbert.

The change in Russell's attitude towards set theory comes out especially in the preface to the second edition of *Principia Mathematica*. Here this powerful philosopher and mathematician draws on the critical analyses of method by Chwistek[67] and Wittgenstein, the latter of which he regards as remarkable enough although not definitely accepting them; on p. XIV he says with resignation: "... it seems that the theory of infinite Dedekindian and well-ordered series largely collapses, so that irrationals, and real numbers generally, can no longer be adequately dealt with. Also Cantor's proof that $2^n > n$ breaks down unless n is finite."

The enquiries of Hilbert and his collaborators seem necessarily to lead to the same result. For the 'method of ideal elements', which Hilbert regards as the specific method for establishing the transfinite, is, as we saw, no more than an expedient symbolism that cannot in any way create new domains of knowledge.

Even if Hilbert makes it his goal to master the transfinite from the finite,[68] this can evidently be achieved only if the transfinite is not a domain of its own kind alongside the finite, but only an abbreviated expression for finite relations, a façon de parler, as he himself has said following Gauss' dictum quoted above. This assertion, however, holds only for the denumerable infinite; the non-denumerable infinite cannot be finitized in this manner.

Let us conclude with a remark of Hilbert's that fixes his finitist conviction so unambiguously that even Kronecker could not have put it more strikingly. "The infinite is nowhere realized, it is neither present in nature, nor admissible as basis for our rational thinking . . ."[69] It is only that Kronecker would doubtless have drawn the consequence that even the supreme master of mathematical method could not succeed in saving Cantor's non-denumerable infinite in any shape or form.

NOTES

[1] An excellent introduction into set theory, intelligible even to the non-mathematician, is Fraenkel's text book, quoted earlier. Intended mainly for mathematicians is F. Hausdorff, *Grundzüge der Mengenlehre*, 2nd edition, Leipzig 1927, a large scale work, including special problems and applications of set theory. A brief and fairly simple account that nevertheless preserves all possible rigour is given by K. Grelling, *Mengenlehre*, Math. phys. Bibl. no. 58, Berlin 1924.

[2] Accordingly the proposition 'there are infinitely many things in the world' is meaningless. As to the assertion that the world is (in its spatial and temporal extension) infinite, the position is this: if the concept of 'the world' is defined meaningfully, namely in relation to 'possible experience', one recognizes that there can be no question of an 'actually infinite' world. If one nevertheless speaks of an 'infinite' world, this can mean only that no definite bounds to extension are being assumed. In the physical world picture of general relativity theory, by contrast, the world has a definite volume; accordingly Einstein's 'world' can be denoted as spatially finite, as against the 'infinite world' of Newton.

[3] About Cantor's definition of a set, cf. above p. 79f. For pure set theory, his following works are fundamental: *Grundlagen einer allgemeinen Mannichfaltigkeitslehre*, Leipzig 1883; 'Mitteilungen zur Lehre vom Transfiniten', I & II, *Zeitschr. f. Phil. u. phil. Kritik*. **91** (1887), 81–125 and 252–270, and **92** (1888), 240–265; 'Beiträge zur Begründung der transfiniten Mengenlehre', I, *Math. Ann.* **46** (1895) 481–512;; II, *ibid.*, **49** (1897), 207–246.

[4] The first proof of the equivalence theorem was given by F. Bernstein (published in E. Borel, *Leçons sur la théorie des fonctions*, Paris 1898). A further proof is due to J. König, 'Zum Kontinuumproblem', *Math. Ann.* **60** (1905), 177–180 and p. 462. However, intuitionists do not accept proofs of this theorem. On elimination of the non-denumerable infinite, it becomes trivial.

[5] *Was sind und was sollen die Zahlen?*, 1887; 4th edition, Brunswick 1918.

[6] An excellent account of the various definitions of finitude is given by A. Tarski, 'Sur les ensembles finis', *Fundamenta Mathematica* **6** (1925). 45–95.

[7] For we can easily show that denumerable sets have the lowest power amongst infinite sets.

[8] Instead of 10, we could equally well adopt the base 2 (binary fractions) or any arbitrary natural number > 1.

[9] Letter to Bernouilli, *Mathematische Schriften* (edited by Gerhardt), III, p. 533.

[10] Published from the literary remains in 1851. Republished by A. Höfler, with annotations by H. Hahn (Philos. Bibl. vol. 99, Leipzig 1920).

[11] It is then easy to show that the concepts of the 'finite' and 'infinite' thus defined coincide with the naive views of the finite and the infinite. Cf. Fraenkel, *Einleitung in die Mengenlehre*, p. 25f.

[12] This description is due to Gauss who used it in a letter to Schumacher in the following context: "... thus I protest ... against the use of an infinite magnitude as something completed, which is never allowed in mathematics. The infinite is merely a 'façon de parler', in that one really speaks of boundaries to which certain relations approach as near as we like while others are allowed to increase without limit" (quoted by Fraenkel, *Einleitung*, p. 1).

[13] Accordingly, one proves by means of the diagonal procedure that alongside algebraic numbers there also exist non-algebraic (transcendental) numbers. This proof is unobjectionable. Further we can give impeccable proofs that every sequence of transcendental numbers determines other transcendental numbers not belonging to that sequence. However, the statement 'the totality of transcendental numbers has a higher power than the totality of algebraic numbers, or, what comes to the same, than the totality of natural numbers' is a formulation that provokes the objections already made and those yet to follow. A general criterion for distinguishing between algebraic and transcendental numbers has not so far been discovered. However, for some particular irrational numbers it has been proved that they are transcendental. The most important are e and π. As is well known, Lindemann's proof that π is transcendental established that the circle cannot be squared by means of ruler and compass alone.

[14] What is meant by the height of an algebraic equation (whose coefficients may be taken as integers without common factor) in this context is the sum of the absolute values of the coefficients and its degree less one.

[15] That is, of rational numbers (cf. above p. 93).

[16] Cf. Poincaré's proof of this theorem in his Göttingen lecture 'Über transfinite Zahlen' (*Sechs Vorträge über ausgewählte Gegenstände aus der reinen Mathematik und mathematischen Physik*, Leipzig and Berlin 1910, p. 45ff.); also Becker, 'Mathematische Existenz', l.c., p. 601, note 2: "What the diagonal procedure shows is strictly speaking this: if we have a denumerable (law-like) series of number sequences, then we can determine, place by place, a number sequence different from all of these".

[17] Cf. these remarks of Brouwer's: "Let us now consider the concept: 'denumerably infinite ordinal number'. From the fact that this concept has a clear and well-defined meaning for both formalist and intuitionist, the former infers the right to create the 'set of all denumerably infinite ordinal numbers', the power of which he calls aleph-one, a right not recognized by the intuitionist. Because it is possible to argue to the satisfaction of both formalist and intuitionist, first, that denumerably infinite sets of denumerably infinite ordinal numbers can be built up in various ways, and second, that for every such set it is possible to assign a denumerably infinite ordinal number not belonging to this set, the formalist concludes: 'aleph-one is greater than aleph-null', a proposition that has no meaning for the intuitionist." ('Intuitionism and Formalism', l.c., p. 91).

The following passage from F. Brentano also deserves mention: "It is one thing to say that each of infinitely many things is consistent, and another that all of them together are. The former is correct, the latter wrong, and by this equivocation many seem to have been the more readily deceived since existence, taken not as consistency but in its primary sense, cannot belong to each without also belonging to the totality." In 'Vom ens rationis', essay from his literary remains, published by Oskar Kraus, Philosoph. Bibl., vol. 193, Leipzig 1925, q. 238ff. (p. 254).

18 Consider the opposite remark by M. Baire (in a letter to M. Hadamard in 1904, quoted by Borel, l.c., p. 152): "As soon as we speak of the infinite (even denumerable, and here I am tempted to be more radical than Borel), the assimilation, conscious or otherwise, to a bag of marbles passing from hand to hand must completely disappear. . ."

19 In place of the covering set (power set) of a given set we can always put the set of all subsets of that set, as a simple argument shows (cf. Fraenkel, l.c., p. 107). Accordingly the observations we have just made hold for that concept too.

20 In *Principia Mathematica* (vol. II, p. 458ff.), as mentioned above on p. 41, the power set is obtained by operating with the 'relation of identity', a procedure devastatingly criticized by Wittgenstein, l.c., 5.4733 and 5.53–5.5352. Following this, cf. also the criticism by Ramsey, l.c., p. 360ff.

21 The associative and commutative laws hold both for addition and multiplication, and the distributive law for the combination of addition and multiplication. A product of cardinal numbers is 0 if and only if at least one factor is 0. The basic rules of calculation with finite numbers can likewise be transferred to potentiation. Thus $m^p \cdot m^q = m^{p+q}$, $m^p \cdot n^p = (mn)^p$, $(m^p)^q = m^{pq}$. However, we cannot form unambiguous inverse operations.

22 The notion of 'preceding', as in the series of natural numbers is to be understood in a non-spatial and non-temporal sense. One term precedes another if the presence of the latter is incompatible with the absence of the former.

23 Cantor has defined operations for calculating both for cardinal numbers and for ordinal types; but for the latter neither addition nor multiplication are commutative.

24 In place of the two principles mentioned we can put the following: if M is a set of ordinal numbers with the property that when any ordinal number appears in M so also does every smaller ordinal number (including 0) then the set M ordered according to the magnitude of the ordinal numbers is well-ordered, and the ordinal number belonging to M is the smallest ordinal number greater than any ordinal number occurring as element in M."

25 "A Theorem Concerning the Inflnite Cardinal Numbers', *Quart. Journ. of Pure and Applied Math.* **35** (1903), 87–94.

26 'Beweis, dass jede Menge wohlgeordnet werden kann', *Math. Ann.* **59** (1904), 514–516; 'Neuer Beweis für die Wohlordnung', *ibid.*, **65** (1908), 107–128.

27 Amongst the most recent enquiries on this problem we must mention above all Hilbert, 'Über das Unendliche'; W. Sierpinski, 'Sur l'hypothèse du continu', *Fundamenta Mathematica* **5** (1924), 177–187.

28 Cf. above p. 81.

29 Cf. W. Ackermann, 'Begründung des "tertium non datur" ', l.c., p. 13ff.

30 We can leave aside an analysis of the finite sense of higher transfinite ordinal numbers, in particular of epsilon numbers, since this is not required for a treatment of the central problem concerning the possibility of an ascent to higher powers in the progression of ordinal numbers.

31 Cf. note 24 above.

[32] Cf. above p. 129.

[33] 'Über Möglichkeiten im Relativkalkül', *Math. Ann.* **76** (1915), 447–470.

[34] 'Logisch-kombinatorische Untersuchungen über die Erfüllbarkeit oder Beweisbarkeit mathematischer Sätze nebst einem Theoreme über dichte Mengen', *Skrifter utgit av Videnskapsselskapet i Kristiania, I., Mathem.-naturw. Klasse*, 1920, no. 4, pp. 1–36. 'Einige Bemerkungen zur axiomatischen Begründung der Mengenlehre', Scientific lectures given at the fifth congress of Scandinavian mathematicians in Helsinki 1922, pp. 217–232, 1923.

[35] Cf. Weyl, 'Über die neue Grundlagenkrise der Mathematik', l.c., p. 46ff.

[36] This was clearly explained by J. v. Neumann, 'Eine Axiomatisierung der Mengenlehre', l.c., p. 229ff.

[37] *Zehn Vorlesungen über die Grundlegung der Mengenlehre*, Leipzig and Berlin 1927, p. 112ff. There he says: "Not only within set theory but in the whole of mathematics, it seems to me that we cannot grasp the non-denumerable infinite in a purely constructive way (namely without enlisting non-predicative processes), unless of course one stretches the concept of 'pure intuition' excessively and insists on thinking of the non-denumerable infinite as immediately given, for example, in the form of the continuum."

[38] Fraenkel, *Einleitung* p. 247.

[39] Cf. above p. 45.

[40] 'Mathematische Existenz', p. 605.

[41] 'Continuous Increasing Functions of Finite and Transfinite Ordinals', *Transact. American Math. Soc.*, **9** (1908), 280–292. Cf. Hausdorff, *Grundzüge der Mengenlehre*, 1st edition p. 114ff.

[42] l.c., p. 795ff. We must however point out that in this context Becker expresses himself so cautiously that it might be more correct to call his thesis a surmise rather than an assertion.

[43] l.c., p. 796.

[44] Against this, Hölder, who sees quite well that the Cantorian continuum cannot be treated arithmetically, wishes it to be regarded as an 'original form' (*Die mathematische Methode*, pp. 349–351). Cf. for example p. 351: "Moreover it may perhaps be appropriate to say of this original form that it is an 'intuition' or that it springs from an ideal intuition." By contrast, cf. our account of mathematical intuition above p. 94ff.

[45] For the history of this problem cf. O. Becker, 'Mathematische Existenz', l.c., p. 569ff.

[46] By contrast, v. Neumann's system of axioms acknowledges the primacy of the theory of well-ordering ('Die Axiomatisierung der Mengenlehre', *Math. Zeitschr.* **27** (1928), 669ff.).

[47] The proof proceeds similarly for any finite integer *n*.

[48] This likewise invalidates arguments like that of F. Bernstein ('Die Mengenlehre Georg Cantors and der Finitismus', *Jahresber. d. Deutschen Math. Ver.* **28** (1919), 63–78, p. 77), who attacks finitist criticisms of classical set theory in terms of the consensus of mathematicians as regards the treatment of set-theoretical problems. Bernstein asserts: "As the strongest general argument against fininitist criticism we have recognized the following: if it is possible that mutually independent enquirers find the same answer to the same question, then we have performed a logical experiment proving that there must be a consistent system of ideas in which the inferences drawn are valid by the strictest standards."

[49] 'Untersuchungen über die Grundlagen der Mengenlehre', I *Math. Ann.* **65** (1908), 261–281.

[50] Besides Fraenkel's axiom system, which we shall presently consider in some detail, we must mention above all that of J. von Neumann, 'Die Axiomatisierung der Mengenlehre', cited in note 46 above.

[51] For what follows, cf. p. 268ff. Fraenkel's axiom system is an improvement of Zermelo's and stands out as being especially clear and simple.

[52] l.c., p. 272. Amongst the *m*, *n*, . . . and so on, we are to understand any objects that are not further specified, between which a relation ϵ likewise not further specified either does or does not hold.

[53] l.c., p. 272f.

[54] l.c., p. 274.

[55] l.c., p. 277.

[56] l.c., p. 278.

[57] l.c., p. 279.

[58] l.c., p. 281. Since in this axiom, whose formulation closely follows Zermelo's, the concept of 'meaningful property' lacks the required precision and led to controversies, Fraenkel essentially tightened it by eliminating that concept and formulating the axiom with the help of the concept of a function. We shall however refrain from giving the reformulation here, because this would require lengthy explanations while being unnecessary for our further remarks.

[59] l.c., p. 283.

[60] l.c., p. 307.

[61] l.c., p. 279.

[62] I owe this point to Carnap.

[63] That the well-ordering theorem follows from the principle of choice is the burden of Zermelo's two proofs of that theorem mentioned in note 26 above. Incidentally, the theorem that arbitrary sets can be compared is likewise logically equivalent to those two principles, as has been proved by F. Hartogs, 'Über das Problem der Wohlordnung', *Math. Ann.* **76** (1915), 438–443.

[64] This short-hand terminology expressed in the usual way is of course not to imply that there is a further mathematical domain beyond the denumerable, which is precisely what we dispute.

[65] This assertion of Hilbert's is matched in Brouwer's terminology by the theorem of the absurdity of the absurdity of the principle of excluded middle.

[66] This is stressed also by Menger in his review of the 3rd edition of Fraenkel's *Einleitung*, in *Monatshefte f. Mathematik u. Physik* **36** (1929), book reviews, p. 7. That classical mathematics too is independent of the concept of the continuum regarded as a set has been pointed out with emphasis by Poincaré even earlier ('Réflexions sur les deux notes précédentes', *Acta Mathematica* **32** (1909), 195–200.

[67] Chwistek has pointed out ('Über die Antinomien der Prinzipien der Mathematik', l.c., p. 243) that Cantor's proof for the existence of the continuum cannot be conducted in the system of Whitehead and Russell leaving aside the axiom of irreducibility, unless with the help of some special hypothesis.

[68] "On the territory of the finite we are thus to furnish free manipulation and full mastery of the transfinite" ('Die logischen Grundlagen der Mathematik', l.c., p. 156).

[69] 'Über das Unendliche', l.c., p. 190. On the same page he says: "The role that remains for the infinite is rather just that of an idea, if by this one understands, in Kant's words, a concept of reason that transcends all experience and by which the concrete is completed in the sense of totality . . ." We would add that this role likewise disappears if one has recognized that the totality itself cannot be conceived precisely.

VI

THE PROBLEM OF COMPLETE DECIDABILITY OF ARITHMETICAL QUESTIONS[1]

The foundations for treating the problem of decidability of arithmetical questions have already been created in earlier sections, namely by the analysis of 'judgments' or 'meaningful assertions', by observing the monomorphism of the axiom systems for the natural numbers and finally by the critical dissolution of the doctrine of the non-denumerable infinite.

Let us start at once with the first of these points: according to it, an assertion is meaningful only if there is a distinguishing feature, simple or composite, in whose presence the assertion is true and in whose absence it is false. Otherwise we have a pseudo-assertion, the illusion of sense usually arising from certain symbols defined for a specific domain being used beyond its boundaries.

Applying this to testing the assertion that a thesis is undecidable from given propositions (axioms) shows that the test must be attended by a criterion of undecidability. Any such criterion for the undecidability of a state of affairs must, however, be replaceable by the proof that it is undecided (that is, objectively undetermined) given the basic assumptions. For let the existence of a state of affairs be objectively determined by the basic assumptions and yet in principle underivable from them, then there would be no criterion for this determinacy; in that case, however, according to the insight formulated above, the assertion of determinacy would have no precise meaning.

After this preliminary remark we proceed to an account of the problem of mathematical decidability.

For the purpose of illustration let us choose the example of Goldbach's hypothesis, whose proof or refutation still eludes mathematicians. The hypothesis runs thus: any even number can be represented as the sum of two prime numbers. For small numbers this has been tested, and a number of approximation theories have been proved.

We now ask ourselves: is it certain that Goldbach's assertion can either be established to hold, or established not to hold? If, however, such a proof or counter-proof be impossible, then the further question arises, whether this impossibility might not reside in an incompleteness of the axiom system, so that we should have to stipulate by an axiom that

Goldbach's proposition held, or that its contradictory held; accordingly we should find instead of the one arithmetic a Goldbachian arithmetic and a non-Goldbachian arithmetic. Moreover, what is good for Goldbach's assertion is good for other unproved assertions, as for instance the famous hypothesis of Fermat.[2] Thus there would likewise have to be a Fermatian arithmetic and non-Fermatian arithmetic. For n propositions assumed to be unprovable and mutually independent in pairs, this would yield 2^n different arithmetics. This is the state of the problem which we shall now discuss.

Let us first clarify under what circumstances one can speak of objective indeterminacy (undecidedness) of a question relatively to certain basic assumptions. A case of such indeterminacy obtains if the question to be decided relates to states of affairs other than those about which the basic assumptions pronounce; as for example in the joke that requires calculation of the captain's age from the dimensions of the ship.[3] This kind of case will hardly need attention as regards the real problems of mathematical axiomatics.

Setting aside this trivial case, there remains (for theoretical questions of the formal kind) only that of polymorphism. For that a state of affairs is not decided by certain judgments can be grounded only in these judgments being unrelated to the state of affairs or in their not sufficiently determining it. A mathematical paradigm for this situation is Euclid's axiom of parallels.

It turns out that either this axiom or its negation can be consistently joined to the other axioms of Hilbert's geometry, so that from these latter we can prove neither Euclid's axiom of parallels nor its contradictory.

This example plays an important part both in the history of mathematics in general and in mathematical axiomatics in particular; and, as is well known, it has provided the main incentive for investigations of the completeness of axiom systems. When doubts arise as to the completeness of arithmetical axiom systems, for example that of Peano, reference is often made to the geometric analogy.

Let us now make clear that the reason for undecidability in geometry does not exist in arithmetic.

In the analysis of the nature of geometrical knowledge we have shown that the sense of the axioms of geometry as given in precise and unintuitive form by Hilbert, lies in the fact that between the three kinds of objects called points, straight lines and planes (the names are of no account as regards the axiomatic system as such), we stipulate logico-mathematical relations. (Thus two points determine a straight line, three points a plane, two straight lines at most one point and so on.) If in this way we stipulate formal relations between the objects, the axiom system can be supplemented as long as certain further such relations between objects remain open. The parallel axiom is just such a supplementation with regard to the other axioms, for they stipulate merely that two straight lines share at most one point, but the division between one point and no point is not as yet effected. Thus, the other axioms do not stipulate how many straight lines through a given point have no point in common with a given line. This 'gap' as to completeness is now filled by Euclid's axiom of parallels or by an axiom determining some non-Euclidean geometry. The indeterminacy (undecidability) of the axiom of parallels with regard to the other axioms thus is due to the fact that they still leave open certain further logical (arithmetical) possibilities.

By contrast, as we have shown in detail in Section III, the definition that we there gave of the number series describes, like Peano's axiom system, the cognitive object natural number and determines it as a logical singularity; and that means that nothing logical pertaining to natural numbers is left open.[4] This axiom system of arithmetic is therefore monomorphic. Thus, as we observed above on p. 60, there can be no branching of arithmetic, with regard, say, to Fermat's theorem or Goldbach's hypothesis, which would correspond to the branching of geometry with regard to the axiom of parallels.

The objective determinacy of arithmetic is hardly in doubt even today, in spite of the fact that many mathematical convictions that were once regarded as unshakeable have been upset, thus producing the foundational crisis. Nevertheless there is a view that might be described as currently prevailing, to the effect that the completeness in the sense of non-branchability of an axiom system is not enough to guarantee its being definitely decidable: even for a non-branchable axiom system it would thus not be certain whether there might not be reasons of principle preventing questions of a certain sort from being solved. The train of thought that leads to this result confronts the infinite domain of numbers with the fact that any proof effected by human thought must remain finite. This is a more stringent formulation of an argument that has played an important role especially in mediaeval philosophy and theology: the finite intellect of man is not capable of grasping the infinite character of the world (of God's idea).

Accordingly, the following possibilities would have to be considered for each unproved arithmetical assertion:[5] (1) the assertion is provable;

(2) the denial of the assertion is provable; (3) the assertion is unprovable and this fact is itself provable; (4) the assertion is unprovable but this fact is itself unprovable.

It is clear that we must most decidedly reject this argumentation. The idea that there are relations that can be grasped only by a finite number of inferences is immediately recognized as nonsense if one has come to see that the infinite itself can be determined in no other way than as the domain of a law. An infinite number of inferences can therefore be meaningfully imagined only in correlation with a governing law of thought that fixes the order of inferences; but this already renders the 'infinite thinking process' finite.

We can now formulate our result as follows: in the concept of complete decidability of arithmetical questions an objective and a subjective (psychological) aspect are merged in a confusing manner. The objective aspect is the unambiguous determinacy of arithmetical objects, namely the monomorphism (non-branchability) of arithmetic as constituted by Peano's axiom system. The subjective aspect lies in the actual capacity or incapacity of human beings to solve problems that are as yet open, where an appropriate empirical assertion would have to give a spatio-temporal boundary for the 'domain of human beings'.[6]

A psychological thesis of undecidability of this and only of this kind might perhaps be put forward for a monomorphic system, but it would be quite irrelevant logically and mathematically: for it is not a theoretical assertion about mathematical relations, but an empirical assertion about our coming to know them in fact.[7]

The connection of the problems concerning definite decidability in arithmetic with those of the infinite in arithmetic stands out particularly well if we consider that statements whose decidability is in question are precisely those that are denoted as transfinite, being erroneously regarded as assertions about a totality of individual states of affairs.

These statements can be divided into two classes: (1) all numbers (or all those that have the 'properties' $p_1, p_2, \ldots, p_n, \ldots$, such as being even, prime and so on) have a further property f; (2) there is a number (or a number that has the 'properties $p_1, p_2, \ldots, p_n, \ldots$) that has the property f'. (Here let us take f' as identical with non-f, so that these two statements are contradictories.)

If, however, we correctly distinguish between specific and individual universality, the absurd notion of running through infinitely many possibilities disappears, and so therefore does the chief motive for the erroneous assumption that there are undecidable arithmetical problems.

Accordingly, the problem of complete decidability of arithmetical questions is most intimately linked with that of the infinite in mathematics in general. Beyond this, there is a special connection between Brouwer's theory of undecidability and the problem of the non-denumerable infinite.

Brouwer himself indeed denotes it as a central point of his doctrine that the negation of definite decidability within the domain of arithmetic is equivalent to the denial of the general validity of the principle of excluded middle.[8] Our previous observations on the latter thesis can therefore be directly transferred to the former.

It thus turns out that the definite decidability of non-denumerable pseudo-domains (such as that of all real numbers) can indeed not be asserted, but the reason for this consists simply and solely in that we cannot logically justify the assumption of such domains.

However, we must clearly distinguish the question here dealt with and answered in the negative (namely whether the assertion of the undecidability in principle of arithmetical problems is compatible with the non-branchability (monomorphism) of arithmetic) from the much more far-reaching 'decision problem'. The latter consists in the combination of the following two problems: 'how can we ascertain, for an arbitrary given expression that does not contain any individual signs, whether the expression does or does not represent a correct assertion for arbitrary insertions for the variables in it?' (the problem of the universal validity of a logical expression). 'Is there any insertion for the variables . . . such that the expression in question represents a correct assertion?' (the problem of satisfiability).

"There two problems are mutually dual. If an expression is not universally valid, then the contradictory can be satisfied, and conversely.

"The decision problem is solved if we know a procedure that allows us, for a given logical expression, to decide by finitely many operations whether it is universally valid, or satisfiable."[9]

As regards the current position of the decision problem, which is the chief problem of mathematical logic, the following remarks are in place:[10] the decision problem has been solved for such problems of the 'restricted functional calculus' (that is, that 'part of the functional calculus' in which there are only such all-statements and existence statements as relate to individuals (counting statements), but not to functional) as contain only functional variables with one argument place; the solution is due to L. Löwenheim[11] and H. Behmann.[12] In the same paper, Löwenheim further shows that in dealing with logical formulae we can restrict ourselves to those in which there are predicate signs with at most two empty places. Special cases of the decision problem for many-place predicates have been solved by P. Bernays and M. Schönfinkel.[13,14]

The most serious difficulties presented by the decision problem are to all appearances those relating to the extended calculus of predicates. Since, however, as we have shown in earlier sections, that calculus must drop away as soon as the correct view of what arithmetic means is given adequate expression in the symbolism of mathematical logic, we may confidently expect a complete solution of this problem in the not too distant future.

At all events our investigations on definite decidability have shown that one cannot prove the decision problem or any other logico-mathematical problem to be unsolvable, nor can one prove that such a problem might be unsolvable yet this fact remain unproved, and so on. This observation is important also from a general epistemological point of view, since it removes the ground from underneath an assertion that has emerged again in recent times, namely that there is in mathematics a kind of being that remains unknowable in principle.

In the next and final section we will show how the antinomies of logic and set theory vanish if we formulate the relevant propositions correctly.

NOTES

[1] By 'arithmetical questions' we understand those that can be defined completely in terms of natural numbers; the whole of analysis thus falls into the domain of arithmetic in this sense.

[2] Cf. above p. 81.

[3] Cf. Geiger, l.c., p. 26.

[4] Cf. what was said about the completeness of axiom systems in Section II.

[5] Cf. P. Lévy, 'Sur le principe du tiers exclu et sur les théorèmes non susceptibles de démonstration', *Revue de Métaphysique et de Morale*, April 1929, quoted in Borel, l.c., p. 265ff.

[6] Cf. above p. 25f.

[7] That monomorphism and definite decidability coincide was maintained already by W. Dubislav ('Über das Verhältnis der Logik zur Mathematik', *Annalen d. Philos.* **1** (1926), p. 193ff., p. 202); also by R. Carnap ('Eigentliche und uneigentliche Begriffe', *Symposion* **1** (1927), 355–374, although in both cases the reasoning is inadequate. Further analyses on this subject appear in a paper by Carnap, still to be published.

[8] Cf. 'Intuitionistische Mengenlehre', l.c., p. 203.

[9] Except for a small syntactic change, the preceding passages in quotation marks are

quoted verbatim from Hilbert and Ackermann, 'Grundzüge der theoretischen Logik', p. 72ff. [E.T., being of later editions, lacks this passage].

[10] Cf. *ibid.*, p. 77ff.

[11] 'Über Möglichkeiten im Relativkalkül', l.c.

[12] 'Beiträge zur Algebra der Logik, insbesondere zum Entscheidungsproblem', *Math. Ann.* **86** (1922), 163–229; cf., also his lectures on 'Entscheidungsproblem und Logik der Beziehungen,' *Jahresbericht d. Deutschen Math. - Ver.* **32** (1923), 66f., and **36** (1927), 17f.

[13] 'Zum Entscheidungsproblem der mathematischen Logik', *Math. Ann.* **99** (1929), 342–372.

[14] During the revision of the proofs for the present book, a further important paper on this subject has appeared: F. P. Ramsey, 'On a Problem of Formal Logic', *Proceedings of the London Math. Soc.* **30** (1929), 264–286.

VII

THE ANTINOMIES

As we mentioned in the introduction to this book, in the last years of the last century, logicians and mathematicians concerned with the theoretical foundations of their science were disturbed by the fact that Cantor's concept of a set and his theory of the actual infinite lead to contradictions.[1]

Let us first consider those antinomies whose formulation rests directly on Cantor's theory of the transfinite. Their dissolution is readily obtained from the results of the proceeding critical analyses.

The oldest set-theoretical antinomy is that of Burali–Forti,[2] the 'paradox of the set of all ordinal numbers'. It arises as follows: consider the set of all ordinal numbers arranged according to magnitude. This set is well-ordered in the sense of Cantor's theory; let us call the set W and the ordinal number corresponding to it ϕ. Then, once more in the sense of Cantor's theory, ϕ is greater than any ordinal number contained in W. This contradicts the assumption that W contains all ordinal numbers. The dissolution of this antinomy is effected by realizing that the concept of the set of all ordinal numbers is meaningless. For, as we have by now sufficiently shown, from being able to make statements about arbitrary ordinal numbers we cannot infer that one can form a self-contained set of 'all' ordinal numbers. The concept of the 'greatest ordinal number' is, however, inconsistent, since for every ordinal number we can, *ex hypothesi*, indicate a next higher ordinal number.

Similar antinomies are those of the 'set of all sets' and the 'set of all cardinal numbers'. The set of all sets would on the one hand have to be the most comprehensive set that contains only sets as its elements, and therefore it would have to have the highest power of all these sets; but on the other hand a well-known theorem of set theory shows that the set of all subsets of this set has a higher power than the original set itself.

Even more closely related with the antinomy of the 'set of all ordinal numbers' is that of the set of all cardinal numbers, which is based on the following two theorems of set theory: (1) For every cardinal number there is a greater cardinal number. (2) The sum of cardinal numbers of a set, which for every cardinal number occurring in it contains a greater such, is greater than any cardinal number of the set.

Accordingly, the sum of all cardinal numbers, which is itself a cardinal number, would have to be greater than any cardinal number, which contains a contradiction.

Concerning the first antinomy the criticism levelled at Burali–Forti's antinomy holds with the additional stricture that in the absence of well-ordering the concept of an 'arbitrary set of sets' is itself meaningless. As to the 'set of all cardinal numbers', we must start by pointing out that we cannot assume a multiplicity of transfinite cardinal numbers, as we have seen. As regards the remaining cardinal numbers we must again stress that they cannot be 'collected into a set'. The concept of 'set of all cardinal numbers' is thus likewise meaningless.

The antinomy called 'Russell's paradox'[3] concerning the 'set of all sets not containing themselves as elements', now to be described, likewise disappears if we eliminate meaningless 'concept formations'.

This antinomy is obtained in the following manner: divide the sets into two disjoint classes such that into the first will go those sets that contain themselves as element, into the second those sets that do not contain themselves as element. Then, presupposing that the principle of comprehension holds, there is a set M whose elements form all and only those sets that do not contain themselves as element; now let us examine whether M falls into the first or the second class.

Suppose first, M contains itself as an element, then M contains an element that contains itself, in contradiction with the hypothesis that M contains only elements not containing themselves. This leaves only the second possibility, that M does not contain itself as element. However, this too leads to a contradiction. For if the set M does not contain itself, then by definition it is an element of M; that is it does contain itself. The concept of a 'set of all sets not containing themselves as elements' is therefore inconsistent.

Russell himself removed this antinomy like the others we have mentioned by his 'vicious circle principle'[4]: "Whatever involves all of a collection must not be one of the collection." Since Russell treats 'set' and propositional function as equivalent, we may put the principle as follows: a propositional function cannot take itself as argument value. On this principle, he bases his simple theory of types, which forms an 'order of types' consisting of 'individuals', 'sets', 'sets of sets' and so on, the mixing of types being excluded. Within the same type Russell does, however, retain the principle of comprehension, since he does not wish to abandon Cantor's doctrine of the non-denumerable infinite.

The 'vicious circle principle' is usually construed as if it created a barrier to thought, which might be expedient for eliminating antinomies but in principle could perhaps be going too far, so that it might perhaps be attenuated without damage to the purpose intended. In fact, however, we saw that no 'set' (whether this word denotes an individual or a specific generality) can contain itself. Thus the 'vicious circle principle' can no more be regarded as a restriction on legitimate thinking than for instance the principle of contradiction; if nevertheless one regards it as a norm, it can be taken only as a norm for the structure of a logically unobjectionable symbolism (language).[5]

Of the antinomies so far treated, Russell's is of special interest to us because it seems to translate readily into purely 'logical' terms; that is, it can be formulated without using the concept of a set. This is done in a manner indicated by Russell, as follows: let a concept be called 'predicable' if it can be asserted of itself, and otherwise 'impredicable'. (Thus for example the concept 'abstract' would be predicable, but the concept 'concrete' impredicable.)

Since the concepts 'predicable' and 'impredicable' are defined as contradictories, every concept would have to be either predicable or impredicable, and no concept could be both. If now we consider the concept 'impredicable' itself, suppose it to be predicable: this would mean that the judgment ' "impredicable" is impredicable' held, which contradicts the hypothesis that ' "impredicable" is predicable'. Therefore our concept would have to be impredicable; but then it could be asserted of itself and so would by definition be predicable.

If we examine this antinomy more closely, we observe that it is not entirely analogous to the 'antinomy of the set of all sets not containing themselves'. The semblance of such an analogy merely arises from the untenable thesis of the principle of comprehension, that with every property we are given a totality of objects having this property. Still they have much in common in that in both cases we find amongst the 'criteria' for the presence of a property P the presence of the property non-P.

The dissolution of this antinomy likewise follows at once, if we reflect that we must distinguish true and false statements from meaningless 'statements', as has been explained more fully in Section II. However, to recognize in individual cases whether we are dealing with a genuine or a pseudo-statement, we must examine what is stated by the proposition in question. That is: we must go back to the criteria for the truth of the proposition, where again we must carefully distinguish between empirical propositions, synthetic *a priori* propositions and tautologies. If we do this in the present case it follows at once that the pseudo-judgment ' "impredicable" is predicable (or impredicable)' is meaningless.

Once we recognize that a meaningful statement is given only if there are criteria for its truth, then antinomies such as that of the lying Cretan (which Russell has reduced to the short form 'I am lying'[6]) likewise vanish. As soon as we ask for the state of affairs that is here asserted as existing we find that there is none, so that we simply have no judgment.

However, behind this sham problem there remains a more deep-seated difficulty which does not become visible until the other has been removed. This is the question whether the object of a judgment can logically include the corresponding act of judging, whether it is possible for a judgment to involve the fact of its being made. Consider the assertion 'all my life I refrain from thinking about my thinking',[7] and assume that the speaker in fact has never thought about his thinking during his whole life, except for the thought expressed in the above judgment. Is this judgment then to be regarded as true or as false? The former would be obtained on the basis of the assumption that the truth or falsehood of a judgment can in no way be influenced by the fact of its having been made. This is the correct formulation of the thesis that a judgment (namely, the object of a judgment) cannot relate to 'itself' (namely to the act of making it). By contrast, the opposite assumption leads us to regard the judgment as false. The analysis of this assumption broaches a central question of psychology, namely the problem of the reflexive nature of thinking.[8] For it seems to belong to the nature of thinking that it can intend itself along with an arbitrary state of affairs.

Whatever might be the outcome of treating this highly important problem, in no case can it entail a logical antinomy. This is recognized as soon as we become aware of the equivocation between act and object of judgment that resides in the term 'judgment'.

The antinomy of the lying Cretan, in its usual formulations, differs from the antinomies considered earlier in that it not only operates with logico-mathematical concepts (or pseudo-concepts), but also contains synthetic concepts. Such antinomies are nowadays called epistemological antinomies,[9] in order to point to the close connection between antinomies and language. For it is usually rooted in the indeterminacy of certain terms in ordinary language, an indeterminacy which, because the expression remains the same while the meaning changes, at first does not appear suspect. In contrast to the epistemological antinomies those discussed earlier are denoted as 'logical' antinomies. The former have not disturbed mathematicians and logicians nearly as much as the latter. Let us now add two of the best known epistemological antinomies.

For every particular natural number there is in a given language, say English, a smallest finite number of signs (including both numerals and letters) by whose combination in linear order this number can be defined. Since, however, with a bounded number of repetitions of a bounded number of signs we can define only finitely many numbers, there must be a smallest number whose definition in English requires at least one thousand signs. Yet the expression of it we have just given defines it in fewer than one thousand signs.

The dissolution of this antinomy follows as soon as we frame the concept of 'definition' correctly. The concept 'smallest number whose expression in English requires at least one thousand signs' is in fact not the definition of a number; for it does not directly relate to the number but only to the thinking of that number, which finds expression in a certain mode of denotation. We might equally well regard the concept 'greatest number written on a certain piece of paper' as the definition of a certain number and then be astonished that this definition no longer coincides with the same number as soon as a longer sequence of numerals than hitherto there recorded is written on that sheet of paper.[10] For with new thoughts there emerge new possibilities of formulation and therefore also of formulation by means of a certain maximum number of signs. Incidentally, if one admitted 'definitions' of numbers by empirical statements about the number signs belonging to them, it would be from the outset impossible to indicate numbers for whose 'definition' one requires at least a certain (sufficiently large) number of signs; for in that case suitably chosen empirical correlations will in general secure independence of the number of signs required for the 'definition' from the magnitude of the number to be defined.

In conclusion let us briefly mention the antinomy of Richard, which is connected with Cantor's diagonal procedure. It arises from the following reflection: it is easy to see that with finitely many signs we can determine unambiguously only denumerably many decimal fractions. For with a given finite number of signs we can determine only finitely many decimal fractions in this way. As a result we can order the finitely definable decimal fractions into a sequence (firstly according to the number of signs required for a definition, and secondly according to the signs used, a serial order or 'alphabet' being presupposed for them). In that case, however, we can use the diagonal procedure to indicate a decimal fraction that does not occur in the above series and should therefore not be finitely definable. However, it has just been defined by the very procedure indicated.

To dissolve this antinomy our remarks about the previous antinomy will suffice. We merely stress that the assumption of concepts not definable in finite terms would be a nonsense.

With this we end our analysis of the logical antinomies.[11] The result is that the seeming contradictions of logic arise from mistakes in thinking and therefore disappear if we avoid making these mistakes. There is a joke that runs as follows: if the result of a calculation does not agree with the tables, then it is always the calculation that is wrong and never the tables. The corresponding thesis for speculations about the consistency of logic is easily derived, and worth taking to heart.

NOTES

[1] Frege was most strongly impressed by this cognitive fact, as is seen from the appendix to his *Grundgesetze der Arithmetik*, vol. II, p. 253. In a letter, Russell had alerted Frege to the 'paradox of the set of all sets not containing themselves' (Russell's paradox, to be analysed shortly). Frege replied that this had shaken the foundations of his theoretical edifice.

[2] 'Una questione sui numeri transfiniti', *Rendiconti del circolo Matematico di Palermo* **11** (1897), 154–164. However, F. Bernstein points out in 'Über die Reihe der transfiniten Ordnungszahlen', *Math. Ann.* **60** (1905), 187–193, that this antinomy had been noticed by Cantor as early as 1895.

[3] Cf. Russell, *The Principles of Mathematics*, Cambridge 1903; Whitehead and Russell, *Principia Mathematica*, vol. I, p. 39ff.

[4] So called "because it enables us to avoid the vicious circles involved in the assumption of illegitimate totalities", *Principia Mathematica*, vol. I, p. 40.

[5] Cf. Wittgenstein, op. cit., 3.332: "No proposition can make a statement about itself, because a propositional sign cannot be contained in itself (that is the whole 'theory of types')."

[6] Incidentally, this statement does not amount to a precise formulation of the kind of antinomy intended, namely that from assuming the truth of a statement its falsehood follows, and from assuming its falsehood its truth follows. For 'I am lying' means 'I knowingly assert something that I consider to be false'; but the fact thus asserted need not on this account be false. The correct formulation of the antinomy to be expressed is this: 'I am uttering a false assertion'.

[7] If formulated with complete precision, this assertion would have to run thus: 'Amongst my thoughts during my whole life there is no thought about my thinking.'

[8] Cf. especially M. Geiger, 'Fragment über den Begriff des Unbewussten und die psychische Realität,' *Jahrb. f. Philosophie und phänomenologische Forschung* **4** (1921), 1–137, particularly p. 44f.

[9] Thus for example Ramsey and Fraenkel.

[10] Cf. also H. Dingler, 'Über die axiomatische Grundlegung der Lehre vom Ding', *Jahresber. d. Deutsch. Math. – Ver.* **28** (1919), 138–158, p. 153ff.

[11] As regards the extensive literature on the problem of the antinomies, see the bibliography in Fraenkel's *Einleitung*. Amongst the most recent contributions to this subject, H. Behmann deserves special attention for his lecture 'Zu den Widersprüchen der Logik und Mengenlehre', given at the annual conference of the Deutsche Math. – Vereinigung, in September 1929.

REMARKS ON THE CONTROVERSY ABOUT THE FOUNDATIONS OF LOGIC AND MATHEMATICS*

The central problem in the theory of a science lies in clearly grasping the sense of the relevant propositions. This process includes: (1) determining the object of the science, for a science consists of statements (judgments) and we cannot grasp a statement if we do not know what it is about; (2) determining the procedure (method) of the science, for we cannot understand a statement if we do not know what it asserts of its object. To observe this, however, amounts to determining what criteria there are for its truth or falsehood, and testing the relevant assumptions in terms of relevant criteria is precisely the procedure (method) of the science.

Once object and procedure are determined, the boundaries of the science have been fixed; indeed, limitation as regards the outside is merely another formulation of definition. For the latter states that science S_1 embraces those and only those statements about a domain of objects O_1, whose verification is effected by means of a procedure P_1; while the former states that statements not relating to the object O_1 and not verifiable by procedure P_1 belong to some science other than S_1.

On closer reflection about the problem area its own possibility begins to look problematic, because of the following consideration: formulating a proposition implies that we associate a definite sense with certain signs, and to understand the proposition is precisely to grasp that sense; what room is left for further problems in which the determination of this sense would be put in question when in fact it has to be presupposed?

The solution of this puzzle lies in the cognitive fact that thinking occurs at different levels of clarity; it turns out that the same thing (obviously the same knowable thing) can be meant more or less vaguely or clearly. The clarification of thoughts in reflection is the task of philosophic thought proper and beyond this of logical and mathematical thought, as we shall see more clearly presently.

The imprecision of thought can lie on the one hand in that we fail to pay attention to certain features belonging to the object[1] in question, and on the other hand in that we link with it certain features alien to it. This concerns above all the 'accidental' empirical setting of objects in the

* First published in German in *Erkenntnis* 2 (1931), 754–779.

contexts of our experience, which misleads us into identifying the objects with the experience. Thus it has nothing to do with the colour red itself that a certain room, which is linked in certain ways with the experiences of my childhood, was papered in red; and it has nothing to do with the nature of the number 90 that it is the number of my house. A primitive man who can count only by assigning the objects to be counted to the fingers of both hands, mentally merges the numbers with his fingers, while a modern man of little education will mostly link them with the conventional number signs, the figures; the mathematician, on the other hand, knows that the laws of numbers are independent not only of the arbitrarily chosen figural aspects of the numerals assigned to the numbers, but also of the fact of the decimal system or indeed of the circumstance whether the grouping of numbers occurs in power series or not. The mathematician is therefore clear in his mind that all these aspects must be excluded if the sense of the number concept or of number theory is to be clarified; for by recognizing the 'foreign bodies' we have already indicated the way for eliminating them.

As regards the example just mentioned, which is important for what follows, one obtains this result: any determinations, with regard to whose change the mathematical calculus remains invariant, do not touch the sense of this calculus and, for the sake of simplicity, are therefore to be eliminated from foundational research.[2]

In particular any symbolic abbreviation is in principle irrelevant in the sense just described, however important it may be as regards heuristic technique. For every symbolic abbreviation is fixed by a nominal definition and must therefore be replaceable by the definiens. If in an individual case such a reduction is impossible, this is a sure sign of a misuse of the symbolic abbreviation, that is a failure to link with it that sense which belongs to it by the definition. In mathematics such abbreviations are found in the case of the so-called 'extensions of the number concept', namely the introduction of negative, fractional, irrational and complex numbers, since all statements about these 'new numbers' are merely symbolically (linguistically) transformed statements about natural numbers; the same holds of the 'ideals' in number theory.

An important and perhaps the most important step towards clarifying the foundational problems of logic and mathematics (or any other science) has been taken when we eschew any 'extravagant' interpretation (going beyond the scope of the definition) of symbolic abbreviations, while exposing within the domain of our enquiry those points at which such a misinterpretation has occurred. A rigorous critique of method in the sense of this maxim leads to the elimination of sham problems.

The foregoing considerations already contain the basic ideas of the analyses that follow; they start from a critique of the current view as to the nature of abstraction.

In the problem of abstraction lies one of the main roots of the problem area of the theory of science in general, and a misconstrual of the nature of abstraction is thus a main starting point for sham problems in the various sciences. As soon as one sees this clearly, a fair number of 'problems' will be recognized as senseless and this removes the obstacles that they have created for legitimate scientific enquiry.

The basic insight that leads to a grasp of abstraction is that the aspects from which we abstract are not eliminated but left open within a certain domain, that is they are assumed to be variable.[3] Let us analyse the abstraction that we perform when we speak of a certain determinate colour. The process is twofold. First we leave open the spatio-temporal determination; that is, whoever makes a statement about a definite colour, asserts what seems to him to hold wherever and whenever such a colour actually exists. Secondly, we leave open the closer determination of those aspects with which the colour appears necessarily linked within the connected unity that constitutes the physical body. Abstraction thus here rests on the invariance of certain aspects within that connected unity with respect to changes in other aspects. Bodies of any shape can be red, blue or green; bodies of any colour can be pyramidal, cylindrical or spherical, but shape without colour or colour without shape are unthinkable.

From the first of these aspects of abstraction we infer a most important consequence, basic for our later enquiries, namely that the result of abstraction (that is, the sense of the concept obtained by abstraction) is independent of where and when the objects exist that individuate the abstracted concept, for example those that possess a certain colour. There is thus no logical correlation between a property and a certain number of objects that have it. To this point we shall repeatedly return, but let us first continue our theoretical analysis of abstraction.

We must, then, ascertain what is the sense of judgments about abstract objects (for example sensible qualities). Here we have to distinguish between empirical and non-empirical judgments.

As an example of a non-empirical judgment take the proposition 'every colour has a certain brightness and a certain degree of saturation'. Here the abstraction rests on the possibility of fixing a certain being-so (an 'existential feature') in thought and to recognize it afresh as the same in various instances (individuations), so that the truth of such assertions remains invariant to the changing empirical circumstances of the relevant observational facts.

As for empirical statements about sensible qualities, they rest on the assumption (basic to any empirical science) that similarities in empirical connections correspond to morphological similarities. Here then the invariance that constitutes abstraction lies in laws of nature, understood sufficiently widely to include pre-scientific rules.

The foregoing considerations show above all that sensible qualities are not independent entities 'alongside' or 'above' actual things in which they exist, but that in making statements about sensible qualities we make assertions about otherwise arbitrary physical bodies that have these qualities. Propositions in which 'sensible qualities' occur as subjects are thus only terminological abbreviations of propositions in which 'arbitrary objects having this quality' are subjects, so that they must be transformable ('translatable back') into such propositions without loss of meaning.

From this it further follows that empirical propositions about properties or relations of sensible qualities are not to be put alongside propositions about objects, but merely represent terminological abbreviations of these. For example the proposition 'the colour green is pleasant to the observer' is to be translated into 'green objects are pleasant to the observer'.[4] Hence a symbolic representation that is to render the structure of states of affairs adequately must either ignore such abbreviations or exhibit them explicitly as abbreviations by indicating the rules for translation into unabbreviated propositions.

What has just been said about sensible qualities holds quite in general of arbitrary 'properties' and 'relations'; wherever there is talk of 'properties of properties', 'properties of relations', 'relations of properties', 'relations of relations', we have merely linguistic abbreviations that have no logical structure of their own kind. It is thus not a property of a definite yellow (exhibited by a definite object) to have a certain brightness, but it is a property of the object to have a colour of that yellow hue and brightness. Notice how what seems to be a relation of subordination from colour to brightness turns out to be a relation of co-ordination of hue and brightness. For in analogous manner we shall recognize in our analyses of mathematical foundations that the seeming functions of functions, functions of functions of functions and so on, are merely functions of two or more variables.

In this connection we must mention especially the 'logical properties of relations' such as symmetry or transitivity. Thus one says that the relation 'neighbour' has the logical property of symmetry, because the proposition '*A* is a neighbour of *B*' means the same as the proposition '*B* is a neighbour of *A*'. Quite apart from our earlier account of the reduction of propositions about properties (relations) to propositions about things that have these properties (relations), we easily see here that it is logically incorrect to interpret the fact that a relation can be linguistically formulated in two ways as a 'property' of that relation.[5]

We shall now analyse the process of abstraction leading to the concepts of 'truth', 'judgment', 'concept', since the insight to be gained is important not only for epistemology in general but also for the specific foundational problems that concern us here.

Let us begin with 'truth'. 'A certain judgment is true' means 'whoever at whatever time makes this judgment, judges truly (in accordance with the facts); he asserts what is the case.'[6] (The criteria by which correctness is determined in each case may here be left out of account.) Thus the concept of the truth of judgments expresses the invariance of their truth with respect to changes in the factual circumstances of the judgment, to wit the person who judges and the place where he judges.

The state of affairs represented was now reinterpreted as if alongside the realm of being there were a 'domain of truth', and a judgment were true if it fell into that domain. However, this insertion is unjustified, as we have just observed; there is no domain of validity to connect being and thinking. 'Judgments as such' and 'truth as such' do not stand beyond or 'above' the psycho-physical subjects and objects about which we judge; rather, these terms are to emphasize invariance to changes in the person who judges and the spatio-temporal data of the act of judgment.

What has just been said has made clear the meaning of 'judgments in the logical sense'. An act of judgment is the thinking of a state of affairs as existing; if now we take account only of that state of affairs (the object of the judgment) and the characteristics by which it is thought (content of the judgment), while abstracting from 'occasional data', namely who judges and where and when, then, if we presuppose perfectly clear thinking, we obtain the 'judgment in the logical sense'.

In the unabbreviated formulations of propositions about the 'truth of judgments' the judgment (in the logical sense) is thus no longer the subject nor 'truth' the predicate; for they run as follows: 'any arbitrary person, at whatever place and time, judges correctly if he judges in a certain way'.

Thus one leaves open the aspects connected with the fact that a judgment occurs, so that a judgment about a judgment in the logical sense can relate only to the state of affairs intended in the latter judgment. In other words: a judgment J_2 about a judgment (in the logical sense) J_1 can relate only to the truth-value of J_1. As we have already shown, however, statements about the truth (falsehood) of judgments are merely rewritings of statements about the being (not-being) of states of affairs.

The process of abstraction resulting in the concept of 'concept' is similar to that leading to the concept of 'judgment in the logical sense'. The concept of an object (state of affairs) is the clear thinking (intending) of this object (state of affairs), leaving open who thinks it and when or where. The concept of a concept is thus a thinking about thinking an object (state of affairs) and so on to any iterated level of reflection. The unabbreviated rendering of these propositions, however, always makes persons figure as subjects and acts of thinking as predicates. Operating with 'concepts of concepts' is analogously a linguistic abbreviation, like operating with 'properties of properties' and 'judgments about judgments'. Very much the same goes for 'predicates of predicates' and 'functions of functions'.

We will now show what disastrous consequences for the theory of logic and mathematics have resulted from a misappraisal of the sense of abstraction and of terminological abbreviations.

In this we may conveniently start from the analysis of the concept of 'logical extension'. Classical logic defines its meaning by the following 'definitions in use':

1. The statement that the extension of the concept C_1 is equal to that of the concept C_2 states that in clear thinking of the domain of objects belonging to C_1 the domain of objects belonging to C_2 is intended as well, and conversely.
2. The statement that the extension of the concept C_1 is greater than that of the concept C_2 or, equivalently, that of C_2 smaller than that of C_1, states that in clear thinking of the domain of objects belonging to C_1 the domain of objects belonging to C_2 is intended as well, but not conversely.

If therefore $E_{c1} > E_{c2}$, then for any arbitrary real (that is, finite) domain the number of its objects that belong to C_2 cannot be greater than the number of its objects belonging to C_1, but these numbers may be equal, if either the domain contains no objects falling under C_1 or it contains only such objects falling under C_1 as also fall under C_2. Thus to a greater extension of a concept there need not correspond a greater number of objects falling under it.

This discrepancy between the inequality relations for logical and numerical extension respectively should in itself be a warning against merging these two 'extensions', but we can infer that this is inadmissible directly from our considerations of principle as to terminological abbreviations and the fact that they can be translated back.

If with this in mind we make it clear to ourselves what is meant by 'equal logical extension' and 'greater (smaller) logical extension' we recognize that this concept is here fixed in such a way that the question as to the size 'as such' of a concept's extension appears senseless. Indeed, we cannot even 'compare' these sizes in the case of any two arbitrary concepts, but only if the meaning of the one is contained in that of the other. This 'meaning relation' does not involve the question how many objects in a certain domain or 'in the whole world' fall under the one and the other concept (the latter question is indeed senseless, since we can make no statements about the world as a whole); hence the logical relation between the extensions of concepts is invariant with respect to all changes in the course of events. Therefore it is incorrect, or misses the intended sense, to define a concept or its extension by the set of its exemplars.

This definition suffers from the mistaken view of the sensualist theory of abstraction, whose question-begging approach it tries to avoid without eliminating the basic mistake of failing to recognize the invariance of a concept's meaning with respect to empirical change.[7]

This criticism strikes not only at the older extensional logic developed by Boole and Schröder but also at modern symbolic logic, and especially at Russell's theory of propositional functions (which, because of its importance, we shall examine in some detail). That theory likewise fails to avoid the mistake of substituting for a concept's universality of kind a numerical universality of the objects falling under that concept. Let us show this very briefly with reference to the account given by R. Carnap with exemplary clarity.

A propositional function is defined as a 'logistic representation of a concept' (as something that can be stated of an object). Propositional functions of one argument represent properties, and those with two or more arguments relations. Thus, to the statement 'the Matterhorn is a mountain in Europe' there corresponds the propositional function 'is a mountain in Europe' or, more usually, 'x is a mountain in Europe'.[8]

The concept of class is now defined from that of universal implication between propositional functions, as follows: if two propositional functions are so related that every object that satisfies the one also satisfies the other, then we say that the first 'universally implies' the second. If two statements mutually stand in this relation, they are called 'universally equivalent' or 'of equal extension'. By means of this definition the concept of extension of a propositional function is introduced, as a concept of a 'quasi-object' (in Carnap's sense); or, what means the same, the sign for extension is thus fixed as an incomplete symbol and the above definition determines (delimits) its legitimate use. A 'class' is then defined as the extension of a propositional function with one argument place.

So far everything seems to be in good order. Nevertheless, this formulation already contains the seed of the misappraisal here under criticism. To grasp this we must make clear to ourselves what is the sense of asserting that two propositional functions are universally equivalent, while being aware that the sense of an assertion is simply the set of criteria whose holding must make us regard the assertion as true; in other words, the method of its verification.

Here we recognize that there are two possibilities of verification, which we shall illustrate by Carnap's example of the two propositional functions '*x* is a man' and '*x* is a featherless biped'.

The first possibility consists in the two concepts 'man' and 'featherless biped' having the same meaning by definition. This is the case either if 'man' is explicitly defined as 'featherless biped' or if the analysis of the two concepts reveals the implicit sameness of their meanings.

The second possibility is to let all men and featherless bipeds pass in review and to observe that each man is a featherless biped and vice-versa. We must of course note that this method of verification can be carried out only in a bounded spatio-temporal domain, from which it further follows that a proposition aiming at empirical verification must fix such a domain. In that case verification is effected by running through the whole domain in a certain order. That order is simply the order of counting.

On the other hand we must note that where equivalence exists in virtue of sameness of meaning, the idea of empirically verifying the equivalence would evidently be senseless.

From this it clearly follows, as could in any case be seen in terms of the identity of 'sense' and 'method of verification', that the two ways of observing 'equivalence' concern two different cognitive objects. In the one case 'equivalence' consists in the sameness of meaning of two propositions, in the other in a certain set of empirical circumstances.

Accordingly, the term 'class' is likewise ambiguous and in statements about classes we must carefully notice which of the two meanings is involved in each case. If, following an example of Carnap's,[9] we start from the proposition 'a certain wall consists of 100 stones' and form the propositional functions 'x is a sub-class of all stones' and 'x has the cardinal number 100', we recognize that the first of the two functions aims at a logical connection of concepts and the second at an empirical observation about stones in a certain wall. This shows itself at once if we form the two propositions 'every stone in this wall is a stone' and 'this wall consists of 100 stones'.

Once this difference is grasped, one recognizes immediately that the empirical fact that certain objects which satisfy one of two propositional functions also satisfy the other can have no significance for the foundations of the non-empirical sciences of logic and mathematics. What is in question here is only the concept of class that derives from sameness of meaning.

However, this 'class' has no fixed number of objects falling under it, and as further emerges (both from our earlier observations on 'properties of properties' and so on, and directly from a clarification of the underlying definition) it is here quite senseless to iterate the concept of class and speak of 'classes of classes'. This has immediate consequences for judging Russell's definition of cardinal number, which we shall have to consider shortly.

First, however, we must explain the closely related ambiguity of the concepts 'all' and 'there is' in the logistic symbolism.

Let us analyse the concept 'all' and illustrate its ambiguity in the following examples: (1) All colours have a certain brightness and degree of saturation; (2) All men are between 40 cm and 2½ m tall.

In proposition (1) the word 'all' says that in the concept of colour brightness and saturation are already implicitly intended; therefore this proposition need not be verified by testing instances, which would in fact be excluded by the absence of boundaries to the spatio-temporal domain that the test should cover. However, even where the meaning of (1) extended only to all the coloured objects in a definite spatio-temporal region, say a certain room, an attempt at such verification would be senseless; for where there is no brightness or degree of saturation, one simply would not be talking of colour.

In proposition (2) we have to begin by clarifying whether one is thinking the property 'height between 40 cm and 2½ m' as contained in meaning within the concept 'man'. If it is, then 'all' does of course mean the same here as in (1); but if (2) is intended as an empirical proposition, then the term 'all' means something quite different.

To see this we must first grasp clearly that an empirical universal proposition for which no area of individuation is indicated remains an incomplete proposition. In order to turn our proposition into an empirically meaningful (that is, in principle empirically verifiable) statement, we must complete it by indicating a spatio-temporal domain,[10] however large, within which anyone can in principle be tested as to his height. Since the meaning of a judgment coincides with the criteria for its verification, let us seek to grasp the meaning that here belongs to the word 'all' by examining under what circumstances our universal proposition can count as verified. This is evidently the case if and only if the following propositions hold: (i) $M_1, M_2 \ldots M_n$ are between 40 cm and $2\frac{1}{2}$ m tall; (ii) There are no other men in the area of individuation concerned (which is in principle empirically ascertainable). In this connection, where all we are after is to focus on the difference between empirical and non-empirical universal propositions, we need not consider questions concerning proposition (ii), which seems (but only seems) to contain a transfinite aspect, but may content ourselves with the observation that, if viewed as empirical, that proposition certainly contains a string of conjunctions; a non-empirical universal proposition cannot be discussed in this way and therefore the concept 'all' has a different meaning here.

Like the term 'all', the term 'there is' has a different meaning depending on whether the existential proposition is empirical or non-empirical. Just as empirical universal statements contain a chain of conjunctions, so empirical existential statements contain a chain of disjunctions. If, for example, a certain domain embraces exactly n objects ($O_1, O_2 \ldots O_n$) with the property P_1 and we assert 'there exists in this domain an object with the properties P_1 and P_2', this proposition is true if O_1 or O_2 or ... O_n have P_2.[11] It is thus an abbreviated formulation for this chain of disjunctions. In non-empirical existential propositions such a dissection is excluded, so that as with the word 'all' we infer that the term 'there is' covers two different concepts. This difference is not abolished by obscuring it through forming the non-concept of 'infinitely many disjunctions' in the case of non-empirical universal propositions: this device serves merely to create confusion.

Let us now apply these considerations to mathematics and ascertain what abstractions underlie the definition of the concept of 'natural number', the central concept of mathematics. It will be expedient to start by analysing the counting process. We count objects by successively attaching signs to them in thought, in an unambiguous way; it being irrelevant to the result who does the counting and when or where, what kind of objects is being counted and what kind of signs used, provided only that the order of use for the signs is unambiguously fixed. If, for example, the signs *a*, *b*, *c* are attached in the order of use determined by the order of writing them down, or the signs *m*, *n*, *p*, then the result of the counting process will be regarded as the same whether *c* or *p* is the sign of the last object counted. For that is the only thing that matters, the sign of the last object counted, here *c* or *p*. Here we come to a new invariance, which underlies the process of abstraction that leads to the concept of natural number: the sign of the last object counted for fixed signs and order of their use, remains the same, whatever the order in which the individual objects are counted (that is, whichever of these signs is attached to any particular object). This describes the abstraction that leads from ordinal number (first, second, third . . .) to cardinal number (1, 2, 3 . . .). The statement that a set of *n* objects is being counted says precisely that however these objects are ordered, the last one will always be the *n*th.

From this we can draw two important inferences: firstly, the question whether cardinal or ordinal number has priority is settled as soon as this 'earlier' is unambiguously described. If we start from the data of experience in order to reach the concept of number by progressive abstraction, the road to cardinal number goes by way of ordinal number and in this sense ordinals appear to come first; if, however, we regard the unity of the set, the invariant aspects within a range, as prior to the variations, then it is cardinal number that comes before ordinal.

Secondly, it emerges that in order to define the concept of number we do not need that of set; the contrary appearance arises from the ambiguity of the word 'set', as we shall presently clarify.

'Set' is regarded as synonymous with 'class' and appears with this meaning as the correlate of 'property'. For one defines the extension of the concept of a certain property as the set of things falling under it. The rejection of this definition emerges directly from what was said above.

Let us once more state that result: we can speak of a totality of real things falling under a certain property (supposing these to be actually distinct individual things) only if a framework (domain of variation) is fixed by means of a spatio-temporal boundary. In that case we can, by means of a suitable 'order of counting' (running through), indicate the number *n* of objects having the given property within the bounded domain, a result that can be expressed by the proposition 'the set of things having the property *P* in domain *D* has *n* elements'. Another, simpler formulation is 'in domain D there are n things having the property P'. The state of affairs in question would, however, be incorrectly described by the proposition 'within domain D the set of all objects having the property P has a sub-set of n elements', for this statement would operate with the totality of all objects having the property P, which in view of what has been said is inadmissible. Thus we cannot assume a co-ordination between a property P as such and a number n, as the set of things having this property; this co-ordination exists only between a property and a specified domain of individuation on one hand and this number on the other.

Even if there is thus no correlation between 'property' and 'set of all things having this property', we might still perhaps indicate a correlation between a special property (kind) and the set of corresponding specifications with a more general framework (genus), above all if we enlist the ultimate objective specifications, namely Husserl's eidetic singularities. If, for example, we consider a certain shade of red (that is, a red such as can be perceived in a particular object), we can ask whether to its concept one might not co-ordinate a 'set of all shades of red'. This calls for the following observations: if we understand the question as asking for the total number of red things with different shades of red (so that for each really specifiable shade we give exactly one representative), then this query is subsumed under the question just analysed, concerning the set of all things having a certain property; not so, however, if independently of all empirical sets of circumstances, we wish to comprehend the range of variation of shades or red as a whole. In that case we find that, since we lack an order of variation attaching a successor to each element (except to the last, if any), this is impossible.

In sum we can say: if we understand 'set' as a synonym for 'property' (or 'propositional function'), then a set does not specify a number of distinct objects belonging to that set (having the property, satisfying the propositional function); if, however, by set we mean number as well, then we also mean a counting order for the elements of that set. The sharp separation of these two meanings of the word 'set' (which corresponds to E. Husserl's distinction between individual and specific universality) opens a safe path through the labyrinth of foundational problems in mathematics. To appreciate this we must now further analyse the concept of cardinal number.

As regards the connection between cardinal and ordinal number the following point remains unclear: counting is a process in time, at any stage the last element counted, whose sign (ordinal index) determines the cardinal number, is in fact the one counted at the latest time; nevertheless the concept of time evidently does not enter into arithmetic, the theory of cardinal numbers, for when we consider any particular arithmetical proposition, we find no temporal reference in it. This fact of cognition was doubtless one of the main psychological reasons behind attempts to determine number independently of the counting process, by defining it as 'property of a set' or 'class of classes'. Thus Russell, like Frege before him, defines the 'cardinal number' of a class α as the class of all classes having the same power as α. Two classes α and β have the same power if there is a relation R by which their respective elements can be put in one-one correlation. The defectiveness of this definition follows from our analysis of the concept of class given above; for one thing, the criteria for the existence of it are being ignored. We now come to a description of the counting process, which will lead us to the definition of natural number. Counting begins with one (the first) object, followed directly by just one more (the second), followed directly by just one more (the third), in which cognitive process the unique immediate precursor of the third object is the second and of the second the first. We continue with this chain of one-one relations between adjacent precursors and successors until the last object, which has no successor. This completely describes the counting process, for there is no other relation that cuts across those described between adjacent precursors and successors. To pass from the counting process to the series of natural numbers two conceptual operations are required. The first lies in isolating the structure of the counting process by abstraction from the phenomenal aspect of temporal sequence. A simple example will illustrate this. That an object is the third in a counting process, means that before it a second and before that a first were drawn into the counting process, but not other objects beyond these. Leaving aside the phenomenon of temporal succession, we obtain this result: something is a third if it presupposes a second and with that also a first and beyond this nothing else. This last restriction is required because a second and first are likewise presupposed by a fourth, fifth and so on. By adding this restrictive condition we have unambiguously marked the 'third'. The logical structure of our counting process is thus a chain of implications.

The second consideration concerns the object counted last in any such process, that is the one object that lacks a successor. Any definite counting process ends with a definite last object, although one is aware that in principle counting could go further, so that in principle there is no bound to counting.

The series of natural numbers is then just the logical structure (abstraction) of the counting process considered as continued without bound. Accordingly, if instead of a chain of implications we speak of a chain of incompatibility relations, the series is defined as follows.

Natural numbers are the elements of the structure defined by the following stipulations and by those alone:

(1) There is one and only one element with whose presence the absence of no other element is incompatible.

(2) For every element N_1 there is one and only one element N_m with whose presence the absence of N_1 is incompatible, while further the presence of N_m is incompatible with the absence only of those elements different from N_1 and N_m whose absence is also incompatible with the presence of N_1.

(3) The relation between N_1 and N_m determined by (2) is incompatible with the same relation between another element and N_m.

This 'presence' is not to be understood as though the signs were 'present' as such: that a particular natural number is present is to mean only that it is conceived as corresponding to an object according to the above formal rules.

This definition corresponds exactly to the classical five axioms of Peano, if the first element is taken as 0 and Peano's 'implicit definitions' of the concepts '0', 'successor', 'number' are transformed into the explicit definition of the concept 'natural number'. Peano's axioms run as follows:

(1) 0 is a number;

(2) The successor of a number is a number;

(3) No two numbers have the same successor;

(4) 0 is not the successor of any number;

(5) Any property of 0 that also belongs to the successor of any number possessing that property, belongs to all numbers.

Here conditions (1)–(3) of our definition correspond to Peano's axioms (1)–(4), while the exclusion of any other conditions in our definition is equivalent to Peano's fifth axiom, known rather infelicitously as the 'principle of complete induction'.

The latter assertion is recognized as true as soon as we remember the following: that every property of 0 that also belongs to the successor of any number possessing it belongs to all numbers, is merely a paraphrase for any number being defined exclusively by the first element (0) and a chain of successor relations; for in that case and only then does a property invariant to the successor relation hold universally of every following term if it holds for the first term. This further catches the sense of a requirement sometimes given in place of Peano's fifth axiom, namely that any number must be finitely reachable from the first, for an 'infinite series of steps' is merely an incorrect formulation of the fact that in a number series not satisfying the first four of Peano's axioms there appear gaps that cannot be formulated with the help of the successor relation.[12]

Our definition of natural numbers has the following advantage over Peano's formulation: firstly our definition clarifies their connection with the process of counting, which in turn clarifies the epistemological status of the concept of number; secondly the riddles surrounding the much-debated principle of complete induction[13] disappear, thirdly and lastly we can immediately infer the consistency of the individual defining determinations as well as recognize that in their totality they completely describe a structure, that is describe it in such a way as to leave nothing open.

Let us begin with consistency: it is immediately obvious that our conditions (1)–(3) are consistent in pairs, for they stipulate that there is one first term, no last term and a one-one relation between adjacent precursor and successor; each of these assertions relates to something different so that they cannot be mutually inconsistent. As far as I know, nobody has ever seriously considered that there might be inconsistency here (or in the first four of Peano's axioms), but the mistrust that did exist was directed exclusively at the rather unclarified fifth axiom. However, as soon as we grasp that its sense lies in the exclusion of any further determinations, this suspicion must vanish. For since this condition does not relate to the other axioms it cannot be inconsistent with them, and besides, the exclusion of any further determinations turns the consistency of axioms (1)–(4) into the consistency of the structure determined by them alone.

Completeness likewise is easily recognized. For the determinations that there is just one first element and no last one, and that the relation between adjacent precursor and successor is one-one, leave no possibility of variation; since further determinations are excluded, this means that no determination leaves such a possibility, so that only one unique structure satisfies the determinations listed. Precisely in this lies the meaning of completness, which on the basis of several definitions that at first appear different is variously called monomorphism, non-branchability or definiteness of decision. The consistency or completeness of this system of determinations defining the series of natural numbers is simply the consistency or completeness of arithmetic (including analysis), since the latter is completely defined by the concept of the natural number series. In particular this implies that there can be no problem of arithmetic that is in principle undecidable. For undecidability is merely indeterminacy within the presuppositions, as is easily seen if we free ourselves from the non-concept of the possibility of infinitely many steps of demonstration.

Moreover, it is the completeness of the system that allows the 'unrestricted application of the law of excluded middle' in mathematics. L. E. J. Brouwer's criticism of this law might make it appear as though mathematical inference consisted in operating with conventionally fixed tools of thought, which proved fit for some cognitive purposes but not for others and would be used according to their fitness – as in the formulation of physical laws in the 'language' of a certain geometry. If, however, we have grasped that logical inference consists merely in making explicit what is already intended in the premisses, we recognize that the possible results of logical inference are completely determined by the content of the premisses in each case. The admissibility of the use of the law of excluded middle in the system of natural numbers means simply that this system is complete, for this condition is necessary and sufficient for any judgment J_i of the system being subject to the following: it is the case that either J_i is true or not-J_i is true.

For 'existentially given' non-denumerably infinite 'domains' (introduced directly or indirectly by formation of power-sets) this disjunction fails, because here we simply have no construction that encompasses all elements of the 'domain', so that we cannot speak of the unambiguous determination of each element of the domain.[14] The clear understanding of the concept of natural number thus acquired is an essential prerequisite for radically dissolving the sham problems connected with the structure of analysis. For as already mentioned, the principle of formation of the natural numbers implicitly contains the whole of analysis – this is now no longer seriously questioned. The so-called extensions of the number concept are merely terminological abbreviations; in principle all propositions about negative, fractional, irrational and imaginary complex numbers can therefore be 'translated back' into propositions about natural numbers. The only one amongst these concept formations that at present still brings with it serious conceptual difficulties is that of irrational (or real) number: these difficulties arise through circular (impredicative) definitions, through the use of higher level concepts and through the 'introduction' of non-denumerable powers. Without going into the theory of real numbers themselves, let us show that these three closely related mathematical sham problems disappear if we become aware of the ambiguity of the concept of set as exhibited above, and draw the appropriate conclusions from this insight.

In the sense of the view criticized (that to each concept there corresponds a set of exemplars falling under it as its extension), propositions of the following kind are regarded as being about sets:

'The set of all natural numbers contains no sub-set of four numbers a, b, c, $n > 2$ such that $a^n + b^n - c^n = 0$.'

We easily see that the term 'set' can be eliminated here; our proposition will then be formulated thus:

If a, b, c, n are natural numbers and $n > 2$, then $a^n + b^n - c^n \neq 0$.

What leads to the use of the concept of a set here is that a universal implication, that is a logical connection of meanings, is interpreted as a statement about a totality of elements. How dangerous this misinterpretation is stands out clearly when we analyse those formulations that involve an iteration of the term 'set', that is where there is talk of 'sets of sets', 'sets of sets of sets' and so on; for in the sense of the term 'set' (or the associated term 'all') where it expresses a universal implication, there can be no question of iteration: any attempt at translating back here shows the senselessness of such a combination of words. Iteration is meaningful only where the term 'set' marks a numerical (individual) universality, for then there corresponds to it a superposition of such correlations. Thus to the proposition 'the set of sets [1, 2, 3, 6, 8], [4, 5, 6], [7, 9] contains three elements' there corresponds the correlation schema

$$\left\{\left[\frac{1,\ 1,\ 1,\ 1,\ 1}{1,\ 2,\ 3,\ 6,\ 8}\right], \left[\frac{2,\ 2,\ 2}{4,\ 5,\ 6}\right], \left[\frac{3,\ 3}{7,\ 9}\right]\right\}$$

or any of the five other such schemata obtainable from the possible permutations in 'counting the sets'; likewise there can be an arbitrary arrangement of elements 'within the sets counted'. The terms 'finite sets of infinite sets', 'infinite sets of finite sets', 'infinite sets of infinite sets' must be similarly understood, provided that where the infinite intervenes we refrain from thinking of a totality of individual discernible elements; what is 'given' is only a law that unambiguously assigns a certain element to each natural number n. Such an assignment is called a 'sequence', so that a particular sequence is characterized by what elements it contains and by the way they are ordered. We speak of the 'set' of these elements in the case of propositions that leave the order open, that is propositions whose truth is invariant with respect to any re-arrangement of the elements. It is, however, a disastrous error to re-interpret such propositions about infinite sets as if they were about unordered infinite sets. As our previous considerations show, this error arises from an unclear grasp of the nature of abstraction or of the closely related ambiguity of the concept of set. For where statements about sets (all-statements) are merely another way of expressing universal implications (logical connections), no ordering schema is presupposed; but in these cases the term 'set' means one thing, while in cases corresponding to existential propositions it means something else. In the former cases we have pure analysis of form, in the latter numerical ordering of number values satisfying this form (formal condition). Failure to distinguish these two aspects, that is identification of a 'function' with the 'set of its values', is a special case of the unjustified identification of a concept with a set of objects regarded as potentially well-ordered. Here is the starting point for impredicative (circular) concept formations, the introduction of the extended predicate calculus, the formation of non-denumerably infinite powers, that is for all those pathways that lead into the thickets of circular inferences and antinomies.

Circularity lies in the fact that where there is no finite bound for the number of values of a function the range of values of that function can be defined only by a general form, so that this form cannot be defined by that range.

The introduction of higher level concepts (functions of functions, sets of sets) arises from misinterpreting superpositions of numerical correlations (more precisely: one-one correlations between ordered n-tuples of natural numbers as domain and certain numbers as converse domain) as 'sequences of sequences . . . (n times)'; or, if the order of the elements remains open, as sets of sets . . . (n times), functions of functions . . . (n times).

For example, that 'within a sequence of sequences' the fifth element of the third sequence is 2, simply states that conformably to the formation law containing two variables the ordered pair 3,5 is unambiguously correlated with the number 2. This is not a 'choice from an infinite domain' – an unrealizable conception –, but the filling of free places in a form.

However, the concept of formation law here remains rather unclear, so that we still need to define it precisely. For simplicity, let us take the case of a sequence; from what has been said this can be easily generalized. The definition runs as follows: the general term of a sequence is the undetermined value, that is the general form of a function of one variable for arbitrary natural numbers as argument. A simple example: the sequence of even numbers in order of magnitude has the general term $2n$, which is simply the form of the function $y = 2x$ for an arbitrary natural number n as argument.

The general term of the sequence does not, however, contain the structure of the sequence, but what characterizes the sequence is that with each particular insertion in this general form a natural number is unambiguously correlated.[15]

Similarly, the general term of a sequence of sequences . . . in n-fold iteration is the general form of a function of n variables for arbitrarily ordered n-tuples of natural numbers as arguments. This state of affairs can alternatively be described as follows: in a sequence, or sequence of sequences . . . arbitrarily often iterated, we must distinguish between the structure itself and the filling of its free places. The structure is unambiguously determined for any fixed degree of iteration, so that there is only one sequence, or better the sequence, the sequence of sequences, and so on; the multiplicity of different sequences, sequences of sequences and so on is merely a correlate to the different functions which in each case determine the general term and therefore the 'filling' of free places.

If by a certain sequence we now understand a certain general term, it is clear in terms of the foregoing analysis that there is no sense in iterating the term 'sequence' thus understood. However, if by 'sequence' we understand the structure just described, then n-fold iteration of the sequence means the unambiguous correlation of ordered n-tuples of natural numbers and certain numbers, which is the schema of an $n + 1$ place correlation unambiguous as regards the last place. There is thus no question of higher level concepts. Universal propositions, asserting a universal validity for a sequence of sequences (or, if the order of elements remains open, for a set of sets) iterated n times, are verified or refuted by analysing the meaning of a mathematical form with n free places and thus do not relate to a multiplicity of elements; on the other hand, existential propositions, asserting satisfiability within the scope of a sequence of sequences (set of sets) iterated n times, concern insertions into this n-place empty schema. Thus in a correct (unabbreviated) symbolic representation of these 'mathematical facts' there can be no universal or existential sign relating to a function, nor an iteration of such signs.[16]

We now come to the third part, namely the elimination of higher transfinite powers (cardinal numbers). To see clearly here, we must analyse the concept of arbitrary sequence of natural numbers; for in the ascent to higher transfinite cardinal numbers, the first step, without whose logical soundness the whole edifice collapses, is defined as the set of all sequences of natural numbers, and this is regarded as the extension of the concept 'arbitrary sequence of natural numbers'. If by a certain sequence we understand a certain general term, what could correspond to an arbitrary sequence is an arbitrary general term; but for a totality of general terms to belong to this concept would require an ordering schema encompassing every conceivable general term. Such a schema is, however, not defined, indeed none can be found, as follows from other considerations based on the famous Löwenheim–Skolem theorem about the satisfiability of any system of counting statements in a denumerable domain.

The second possibility of defining an arbitrary sequence consists in viewing it as the structure common to all definite sequences; however, to this concept no multiplicity can belong, because, as already observed, it is completely determined and thus offers no room for variation.

It is thus impossible to form a concept of 'arbitrary sequence' of natural numbers in such a way that a multiplicity of sequences including any definite such sequence corresponds to it. However, it is in postulating the existence of such a multiplicity and in nothing else, that the 'procedure' of forming power sets consists, which in set theory is an indispensable prerequisite for any ascent to higher transfinite powers. For it turns out on closer reflection that the apparent alternative of ascent to higher transfinite classes of numbers by means of the theory of well-ordered sets alone does not exist; for this I refer the reader to my earlier book *Das Unendliche in der Mathematik und seine Ausschaltung*, p. 157ff. [E.T., this volume p.128ff.].

For in order to ascend to the smallest class of numbers belonging to non-denumerable sets one already presupposes the existence of non-denumerable powers; but the sole support for this assumption is the pseudo-procedure of forming power sets.[17] However, are not the propositions about the non-denumerably infinite – so the objection to this result will run – used time and again in the most varied branches of mathematics and above all in analysis? How can it be meaningless to talk of non-denumerable sets if we can perform logically correct operations with this concept? This objection is invalidated by the insight that these doubtless legitimate propositions do not concern the 'non-denumerable', but – as in the case of operations with higher level concepts – the appearance that they do is provoked by terminological abbreviations. At the base of this we mostly find the mistake of identifying a mathematical form with the set of the values satisfying it,[18] a view criticized in detail above.

Once we have seen through this state of affairs, we recognize that the acquisitions of classical mathematics remain untouched by the elimination of the non-denumerable, and the same holds for topology, as far as I can see at present. In particular we mention that operations with the concept of 'ordinal continuum' in topology do not presuppose the existence of a power $2^{\aleph_0}$. It remains to examine whether some of the especially difficult questions in current topological enquiries might not turn out to be pseudo-problems.

With the elimination of non-denumerably infinite domains the set-theoretical antinomies likewise disappear, and the same holds for the corresponding logical antinomies in the strict sense, as soon as we see through the inadmissible merging of logical extensions of concepts (classes) with numerical multiplicities. This correspondence is incidentally a confirmation of the thesis argued above, that the formation of non-denumerable powers goes back to incorrect operations with the extensions of concepts (classes). The principle underlying Russell's simple theory of types that 'a propositional function cannot take itself as an argument', removes enough of these mistakes to eliminate the logical set-theoretical antinomies properly so called,[19] but the uneliminated remainder[20] require the introduction of logically unjustifiable axioms of existence (axioms of infinity, reducibility, choice) in order to construct mathematics, which again leads to insuperable difficulties, especially in the theory of real numbers.

Since the basic mistake criticized in this essay is deeply rooted in mathematical and logistic semantics, its eradication in many cases encounters enormous difficulties; but these are of a purely technical nature and their solution therefore is in principle of minor philosophic interest, important though it is for simplifying the decision problem of mathematics.

However this may be, the critique here presented is purely immanent and therefore not tied to a particular point of view; for the practice to which it objects is the ambiguous use of signs, and a refutation of that critique could be seen only in the invalidation of this objection, but not in a reference to technical difficulties.[21]

NOTES

[1] The word 'object' is understood in the widest sense.

[2] Within the framework of mathematics itself there are many applications of this principle. One of the best-known is the principle of duality in projective geometry.

[3] That abstraction cannot mean elimination, can be seen from the following reflection: an act of thinking as such cannot change the object thought in it, but it is inadequate (false) if the object is not thought as it is. Hence an independently considered property ought to be able to exist independently, a consequence in fact drawn by conceptual realists (Plato, Aquinas and others), but untenable in the light of closer analysis. From

this one can easily see how closely much in the history of metaphysics – especially the controversy on the nature of universals – is linked with the problem of abstraction.

[4] We need not here go into the further transformation of this proposition which carries within itself the analysis of the causal relation.

[5] Cf. also Wittgenstein's *Tractatus Logico-Philosophicus*, London 1922, proposition 4.122ff.

[6] It could of course be that the expressions acquire their full sense only by our considering the personal or spatio-temporal data of the fact of their being expressed. This happens above all where personal and demonstrative pronouns appear in the propositions (for example 'you have insulted me', 'the cross-road is not here'). In such cases Husserl speaks of 'essentially occasional expressions' (*Logische Untersuchungen*, II/1, p. 81 [E.T. p. 315]). To decide whether such a proposition is true or false, we must first eliminate the occasional expressions and replace them by the corresponding objective determinations.

[7] This misappraisal is avoided if one sees that logical connection is a connection of meanings. That a judgment J_2 logically follows from a judgment J_1 simply asserts that the judgment J_2 clearly thought contains no assertion not also contained in the judgment J_1 clearly thought. Since, however, the meaning of an assertion is simply the set of criteria for its truth, this means that the method of verification of J_2 is contained in that of J_1. However, the logical connection is completely independent of the hypothesis of the truth of J_1; likewise if J_1 and J_2 are both false, or J_1 false and J_2 true.

[8] *Abriss der Logistik*, Vienna 1929, p. 4.

[9] *Der logische Aufbau der Welt*, Leipzig 1928, p. 51 [E.T. p. 64].

[10] In universal propositions empirically meant a more or less precisely defined domain of individuation is usually implicit. In our example, the earth and the 'present' are more or less precisely intended as spatial and temporal references respectively. This also holds for the formulation of any law of nature, however abstract, in so far as it is a genuine empirical law and not a concealed convention or mere schema for judgments (like the law of causality, for example). The spatial domain of individuation is here understood as the sphere directly or indirectly observable in each case (unless a stricter concept is fixed), and the temporal domain as the 'present' or a period not usually specified within which the present lies.

[11] Non-exclusive disjunctions.

[12] The simplest example are series of the ordinal type $\omega + {}^*\omega + \omega$, for example $-1, -\frac{1}{2}, -\frac{1}{4}, -\frac{1}{8}, \ldots \frac{1}{8}, \frac{1}{4}, \frac{1}{2}, 1, 2, 4. \ldots$

[13] The so-called recursive method, which by means of the principle of complete induction seems to summarize an unlimited chain of inferences, in fact means a mental operation with the general form of the natural numbers. Cf. below p. 183ff. and Wittgenstein, op. cit., proposition 6.022.

[14] That the 'applicability of the law of excluded middle' within a system of statements (in Brouwer's sense) simply means the completeness of this system, is likewise evident from the fact that Brouwer himself calls it a central point of his theory that the negation of the definiteness of decision in arithmetic coincides with the negation of the general validity of the law of excluded middle. (Cf. 'Intuitionistische Mengenlehre', *Jahresber. d. Deutsch. Math.-Ver.* **28** (1919), 203–208.) For definiteness of decision is simply completeness. However, Brouwer leaves the completeness of arithmetic out of account and

accordingly prohibits the use of the law of excluded middle even 'within the denumerable', which creates the false impression that his basic and justified constructivist thesis leads to an impoverishment of classical mathematics.

[15] We presuppose that the function is not undefined for 'almost all' integer argument places. Otherwise no sequence corresponds to it.

[16] As far as first level existential propositions are concerned, I must here content myself with a reference to the relevant analysis in my book *Das Unendliche in der Mathematik und seine Ausschaltung*, Leipzig and Vienna 1930, p. 63 (especially p. 66f.). [E.T., this volume, p. 55ff.]. If its result, that any verification of an existential proposition must implicitly indicate an instance, is regarded as valid, the existential quantifier would fall away as a basic concept. Propositions of the form 'there are numbers having the property *P*' would simply be requests to exhibit such numbers in fact.

[17] It is a widespread error that Cantor's diagonal procedure proves the existence of higher powers. In fact this merely proves that for any arbitrary sequence of sequences one can construct a sequence not contained in it. (The diagonal procedure indicates this construction.) Only if we already presuppose the logical existence of a set of all sequences, which, as we saw, is a mistake, the above proof shows 'that the set of all sequences cannot be represented one-one on the set of natural numbers'. If of course the existence of a non-denumerable power is to mean merely this cognitive fact established by the diagonal procedure, then everything is all right, but this merely introduces a certain terminology; it cannot give any sense to the central problem area of set theory, rooted as it is in a constructive attainment of higher transfinite powers (especially the problem of the continuum).

What can be taken even less seriously is the appeal to the intuitively given continuum. For it is now definitively established that no intuition can relate to an infinite multiplicity, and that appeal to so-called geometric intuition (an epistemological hybrid) in mathematics and its theoretical analysis is inadmissible.

[18] Cf. *Das Unendliche in der Mathematik*, p. 169ff. [E.T. this volume, p. 137ff.]

[19] The so-called epistemological antinomies are not thereby removed; their elimination is carried out by Russell with the help of his ramified theory of types.

[20] As we have recognized, no propositional function can be an argument of any other.

[21] The foregoing enquiry has been conducted at a depth just sufficient for obtaining the results of the critique of method, and for want of space the formulation had to be rather brief. For completeness let me therefore refer to my book (op. cit.) and to an essay shortly to appear in *Jahrbuch für Philosophie und phänomenologische Forschung* (edited by E. Husserl) under the title 'Logische Prinzipienfragen in der mathematischen Grundlagenforschung'. [E.T., this volume, p. 188ff.].

QUESTIONS OF LOGICAL PRINCIPLE IN THE INVESTIGATION OF THE FOUNDATIONS OF MATHEMATICS*

INTRODUCTION

The development of mathematics in the last hundred years and the theoretical reflections to which this has given rise have produced results that seem apt to bring about a complete overturning of traditional conceptions of logic. It is not only the epistemological significance of logical principles that is drawn into the debate in order to elucidate the origins of a validity that is not as such denied – as was for example the case in the controversy over psychologism – but the very principles themselves are put in question. Thus the antinomies that appear in Cantor's theory of manifolds – of which some can be applied to logic in the narrow sense as well – have helped to undermine our conviction that logic is consistent. In the course of the critique of method that was thereby provoked, the universal validity of the principle of excluded middle was challenged and restricted to finite domains.

If this amounts to an attack on basic positions of classical logic, there is on the other hand an elaboration of what are called logical calculi, by Boole, Schröder, Frege, Whitehead, Russell and others, and this seems to have widened the scope of logic so enormously that Kant's much quoted dictum about logic having been unable to make essential progress since Aristotle's time must surely appear outdated.

Indeed, the investigations of these enquirers are designed to show that mathematics contains a series of inferential modes that are alien to syllogism, but that between them and syllogism there is no difference so fundamental as to warrant making it the basis for a break between logical and mathematical method; therefore mathematics must be regarded as a part of logic. The opposite tenet, that of the intuitive evidence of mathematics, in particular of Euclidean geometry, which is over two thousand years old and has received its deepest philosophic foundations in Kant's *Critique of Pure Reason*, had been uprooted by mathematical progress, especially by the discovery of non-Euclidean geometries and the consequent arithmetization of geometry (as well as by the construction of curves that

* Composed in German, not hitherto published, though intended for Husserl's *Jahrbuch für Philosophie und phänomenologische Forschung*. See p. 187 above.

contain all points of a surface and are therefore counter-intuitive). This revealed the heuristic efficacy of a mode of thought that sharply distinguishes between a calculus as such and its intuitive interpretations. Because of the attendant drift towards formalization, this method, far beyond its immediate results, has influenced mathematical thinking (witness for instance F. Klein's *Erlanger Programm*), pointed to new ways in mathematical foundation research and led to new conceptions as to the nature of logic. For this formalizing tendency leads to the axiomatic method, or more correctly, to the new form of that method, for which it is characteristic that in the 'axioms' we operate with undefined basic concepts devoid of all fixed content, so that the individual axioms are schemata of a formal theory which is treated as a calculus without reference to its intuitive models.[1] The axioms, in contrast with those of Euclid, are not propositions but propositional functions that do not yield true or false judgments until the free places have been filled by symbols marking the objects of the various systems.

If in this way we recognize that operations with the axiom system leave questions as to truth or falsehood completely out of account, it seems natural to enquire whether, by axiomatising logic, we might show that in its operations too the concept of truth with its attendant problems is irrelevant.

There is more to come: by systematically developing the axiomatic method, which reaches its peak in Hilbert's formalism, it is hoped to show that not only the question of verification but also every form of meaning-content is alien to the calculus; in other words, that meaning is not immanent in the calculus but only conferred on it, by interpretation.

Let us be clear what this last step on the road to formalization signifies, in view of the stage marked by Hilbert's *Foundations of Geometry*. There, the basic elements 'point', 'straight line', 'plane' are indeed devoid of all intuitive meaning, but between them logico-arithmetic relations are set up so that they are meaningfully combined with each other. Thus to a pair of points there corresponds one straight line, to three points one plane, to two straight lines at most one point and so on. In the calculus of Hilbert's proof theory, however, there are, on the prevailing view (which we will discuss later), only elements that are figures and certain complexes ('formulae') formed from these elements, from which according to certain rules further such complexes are formed from the same elements. By interpreting such 'figures of proof' (by 'inserting' values), we then obtain mathematical proofs, which seems to imply that the conceptual process characteristic of the proof is pre-formed in the stages by which the proof figure was set up. Since the mathematical calculus is the mode of expression or 'language' of mathematics, we reach the further conclusion that *language*, or at least the language of mathematics and of logic in the narrow sense (to which the above comments about mathematics are readily extended) is prior to the attribution of meaning; a result which, if correct, would signify a total collapse of traditional views on the relation between language and thought.

The above remarks are only a survey sketch (without claims to completeness) of the fundamental problems of logic that have arisen within the framework of recent foundational enquiries in mathematics or have been given a new formulation in this context.

That the problems arose as they did explains why attempts at solving them are made mainly, though not exclusively, by mathematicians using mathematical symbolism, or a logistic one that imitates it. The majority of professional philosophers who lack the relevant mathematical expertise, have thus been excluded from testing the results obtained. On the other side, most mathematicians working on foundational problems remain unfamiliar with, or partly fail to appreciate, a series of logico-philosophical writings, which, if properly understood, could have greatly helped them in dealing with their problems.

In spite of this splendid isolation from philosophy, or perhaps just because of it, even the greatest mathematicians when working on foundational problems have not remained free from prejudice in favour of certain speculative positions; accordingly they have misconstrued the insights they have gained by otherwise exemplary analyses of mathematical and logical thought. It is in this irreconcilable clash of interpretations from different points of view, and not in the results produced by descriptive analysis themselves, that we must seek the root of most of the seemingly unbridgeable oppositions between the various orientations within logico-mathematical foundational enquiries, in particular that between formalism and intuitionism.

If we jettison this speculative ballast, careful examination of the problematic position will convince us that the above questions are ready to be solved, but that radical solutions lie at deep levels where they have not often been looked for in the past. The present enquiry is designed to support that thesis. In doing so we shall start from the question how logic and language are related, almost an all-pervasive problem in the relevant field. Next, we shall analyse the sense of logical propositions and principles, which will involve a consideration of the 'antinomies'; and finally with the brevity imposed on us, we shall examine the relation of logic and mathematics.

I. LOGIC AND LANGUAGE

At the beginning of any theoretical analysis of language there must be the reflective insight, that language is not a system of acoustic complexes ('words') and their configurations ('sentences'), but a system of rules that correlate these complexes and configurations with certain contents of thought. It is thus not the acoustic complex 'blue' as such that is a 'word' in English, but that complex in so far as it 'means' that specific contents of experience that is in the first instance accessible to sight. This becomes obvious if we consider that two languages can have all words and grammatical rules in common and yet be different languages, if different meanings are assigned to words (here understood as mere acoustic complexes) and their combinations.

We may here refrain from analysing assigning relations, since they have been admirably analysed in the first and sixth of Husserl's *Logical Investigations*. We therefore confine ourselves to citing a passage from these analyses that is important for our subsequent enquiry, concerning the difference between *image* and *sign*: "Pictorial imagination evidently has the peculiar feature that, wherever its goal is realized, the object appearing to it as image identifies itself by likeness to the object given in the act of realization. To call this a peculiar feature of pictorial imagination amounts to saying that realizing like by like internally determines the character of the realizing synthesis as imaginative. On the other hand, if because of a chance likeness between sign and signified there arises a recognition of their mutual likeness, this does not realize the assigning intention; quite apart from the fact that the recognition altogether lacks that peculiar identifying awareness which makes like coincide with like in the way of image and thing. Rather it is of the essence of a signifying intention that in it the apparent object of the intending act and that of the realizing act (for example name and thing named in their realized unity) have nothing to do with each other."[1]

This at once gives rise to some important consequences. The first that we must point to concerns the 'logical priority' of thought over language, consisting in the fact that the concept of speech already contains that of thought but not conversely. How far thought is normally permeated by assigned attendant conceptions is a question of the empirical psychology of thinking, which remains entirely out of play as regards our problem.

The second consequence concerns the problem of speaking about language. If, wrongly, one regards language as a system of acoustic complexes and their configurations, a purely acoustic phenomenon, then propositions about language figure as statements about these phenomena as such and are therefore obviously quite different from the propositions of the language itself, where what matters is the meaning of the acoustic phenomena. If, however, one has recognized that the concept of a linguistic sign already contains the semantic aspect, then speaking about language appears in quite a different light; for the meanings of acoustic combinations, which alone are the target of the speaker's intention, now form the object of the statement even where we speak 'about language'.

To attain full clarity here, we must digress and examine what is meant by 'utterances about something', a matter of some importance also to other problems that concern us in this essay. The aim is to show that under this heading two disparate concepts have been linked, those, namely, of empirical assertion and of reflective elucidation. For both in ordinary and in scientific language we designate as a statement about a certain spatio-temporal object, say a house, that it has shape and colour, while designating as a statement about colours that they differ in hue, brightness and saturation, or as one about numbers that they are divisible by other numbers. Let us clarify the difference between statements of the first and second kind: if we say of a body specified by its spatio-temporal position that it has a certain property (say, a colour), we make a genuine assertion that can be confirmed or refuted by experience. This thing there is so, but it might be otherwise, too. However, if 'about colours' we make the statement that they can be differentiated as to hue, brightness and saturation or about squares of natural numbers that they are either divisible by 4 or leave a remainder of 1 when divided by 4, we are not saying anything 'new' about the object that figures in the statement; that is, not anything that is not already contained in its concept. It is obvious at once, that the difference here outlined is that between empirical and non-empirical propositions.

We leave open the question whether non-empirical propositions might be divided into analytic and *a priori* synthetic judgments.[2] The main thing for us remains that all non-empirical statements are pure elucidations, that is, phenomenologically speaking, results of a process of clarification. The elucidation of thoughts, however, does not refer directly to the objects of thought, but to our initially confused thinking of objects that is to be clarified by the elucidation. If for example I constate that colours as such can vary as to hue, brightness and saturation, this is not a statement 'about colours' (since the concept of colour, that is our clear thinking of colour, already contains these three degrees of freedom), but a clearer way of thinking about the essence of colour (about colour as such). With the process of clarification as such and with its constitutive presuppositions (which form a central problem of philosophy) we need not here concern ourselves. The problem to be tackled here, as already mentioned, arises from the fact that people mix up the properties of things ('concrete' in the sense of the third of Husserl's *Logical Investigations*)[3] with 'properties' of properties, and likewise relations between things with relations between properties or other 'abstracta'. We shall shortly have to consider the dire consequences flowing from this; for the present it will suffice to remark, entirely in Aristotelian vein, that only a body or person (a physical or psycho-physical concrete object) can be the subject of a statement. This implies in particular that wherever linguistic formulations lead to subjects that are so-called 'higher-order objects' such as society, the state, the economy, this amounts only to an abbreviated mode of expression which must be 'translated back' for clarification.

This insight enables us to formulate clearly what we are to understand by 'speaking about language'. As we saw, language is a system of co-ordinating rules between acoustic complexes and their configurations on one side with thoughts of objects or facts of the world on the other. In view of what we have just said, we cannot speak about these rules, we can only elucidate them. Speaking about language is thus nothing but elucidating language rules.

At this point a mistaken view frequently arises from mixing up the concept of language just mentioned with that of language as a social phenomenon; that is, with the fact that certain groups of people uniformly use a certain acoustic symbolism for communicating with each other.

'Language' as a social fact does indeed enter the content of empirical statements but language here does not figure as the subject 'about' which we speak; for in characterizing those who speak a language one is not speaking 'about that language' (no more than one is speaking about the colour 'yellow' when one observes that three yellow birds are sitting on a certain tree). Rather, when one asserts that they use that language one is making a statement about them.

We can bring this out by the following consideration which is important for the problems concerning the foundation of mathematics and logic.

'Language' as a set of rules co-ordinating acoustic complexes and their configurations on the one hand, and thoughts of objects and facts of the world on the other, is already the product of a process of abstraction; since in this formation of concepts we have abstracted from the specifications of what people use these rules and where and when. It follows that from this concept, which is of its essence indifferent to individuation, it is impossible to derive individual specifications, and likewise for all concepts of abstract objects. This does not mean that we now have to assume two mutually co-ordinated existential domains, one of real and one of ideal objects, for the ideal objects are dependent in Husserl's sense; that is, they are the result of conceptually isolating certain aspects that remain invariant to changes in other aspects basically linked with them.[4] Thus we can speak of colour as such or shape as such – though they are basically linked, so that neither can ever occur without the other – because arbitrary shapes can be linked with arbitrary colours. The aspects from which we abstract are thus not eliminated (how could thought sever what belongs ontically together?), but merely left indefinite, that is, taken as variable. The genuinely phenomenological problem of abstraction then consists in making explicit the cognitive sources that render it possible at all, in the present case the evidential sources for every colour being 'really' compatible with any shape, though it appears in principle impossible to verify such an assertion by trial. It has been doubly fatal for modern empiricism, which arose out of the British tradition of sensualism, that it has not to date seen through this impossibility and thus confuses individual (numerical) with sortal (general) universality.[5] For one thing, the circularity inherent in the empiricist theory of abstraction has introduced absurdities into the theory of logic and mathematics (to be discussed later), and for another this has blunted our vision for grasping general problems and therefore prevented us from seeing those fundamental philosophical questions that are treated by transcendental phenomenology.

After this short digression on the problem of abstraction, let us return to the question of speaking about language. We have already observed that such discourse merely consists in elucidating that language, with the result that we cannot make therein any statements save those already contained in the rules forming the criteria for a certain utterance appearing as an utterance 'in a certain language'. Thus we do not here have to distinguish between two disparate problem areas ('speaking in a language' and 'speaking about that language'); we have to do merely with a change of perspective from natural to reflective outlook (as specified in more detail by Husserl). Whereas in 'speaking in a language' we focus on the objects to which we refer by signs, in 'speaking about language' we focus on the referring relation itself. It is the shift of attitude that is often symbolized in writing by means of quotation marks, to indicate that we are speaking not about the object but about the concept of that object. As will emerge in the sequel, this observation is important for determining the relation between logic and metalogic, and mathematics and metamathematics.[6]

On the basis of our enquiry to date we can now decide the question (important for foundational enquiry in logic and mathematics) whether it is possible to speak in a given language about that self-same language. In view of Russell's theory of types, which bars a propositional function from containing itself as an argument,[7] this possibility has been denied. Our present analysis leads to the following results.

If we see in language a system of certain co-ordination rules of the kind specified above, abstracting from all incidental aspects, then speaking about language is strictly impossible and *a fortiori* so is speaking about a language in that language. There is, however, no obstacle in principle to formulating the rules of a language in terms of the semantic means of expression laid down by these rules themselves, as indeed happens in any single-language grammar book. A statement such as 'adjectives are indeclinable' does not assert about itself that it is a sentence of English (considered as a system of rules), but it is such a sentence, in so far as something is meant by it.

If, however, by a certain language we understand the usual means of communication of a certain group of people, we have, as already observed, statements about how certain people communicate with each other.

In this connection we must keep in mind a further point. We have emphasized that language is not a system of acoustic complexes and their configurations, but a system of co-ordinating rules between these and thoughts of objects and facts in the world. One can of course readily abstract from the fact that a certain acoustic complex means just 'this thing here' or 'something like this', and confine oneself to registering the acoustic complexes that have a meaning and the rules with whose help one forms meaningful configurations. However, the above observation of principle is in no way impaired by this. For it is no more a property of acoustic complexes (and their configurations) as such that they have a certain meaning than it is their property to have a meaning at all; indeed, that they have a meaning amounts to no more than that this meaning (here left unspecified) is linked by speakers with those sounds in that these speakers construe them as symptoms for thought contents.

We can carry abstraction one last step further, by classifying acoustic complexes and configurations as 'meaningful' or 'unmeaningful', where we moreover leave aside the content of the concept of 'meaning', and the classification assigns a certain index, say '*b*', unassociated with any meaning, to certain acoustic complexes and configurations externally (acoustically) characterized, and to these alone. Even at this level of abstraction, which must be conceptually grasped in view of our subsequent discussion of the 'formalist' problem, we readily see that what has been said concerning 'speaking about language' remains applicable. For denoting mere acoustic complexes and their configurations in the manner outlined as *b*-complexes or configurations is not a 'statement about' them but a definition in use of the concepts '*b*-complex or configuration'.

We can thus summarize our results as follows: there is no 'speaking about language' in the sense that language might be an 'object about which' one could speak, as one does about a body or a person; appearances to the contrary vanish if we formulate clearly what is really meant when, as it is imprecisely put, we speak 'about language'.

There can therefore be no hierarchy of languages if by this is meant that for a 'language about language' the latter is an 'object about which . . .'; rather, such a hierarchy can be conceived only as a shift of focus in the way described above, that is as an iteration of levels of reflection. Once this is grasped in principle we can keep the usual mode of expression in which language figures as a subject, to avoid excession complication in what follows.

We can now broach the central problems of the logic of language themselves and begin by clarifying their meaning. This is best done by asking oneself whether our previous definition of language as a system of rules co-ordinating acoustic complexes and their configurations on the one hand with thoughts of objects and facts on the other, was not too wide, because about the manner of forming configurations (propositions) we can make *a priori* constatations that are valid for any arbitrary language and would thus have to enter into the definition of 'language'. This involves discussing the 'idea of pure grammar' as developed in the fourth of Husserl's *Logical Investigations* and since opened up especially in Wittgenstein's *Tractatus Logico-Philosophicus.*[8] To exclude obvious misunderstandings, we must observe the following:

We can constate that a definition is too narrow or too wide or altogether inadequate only by showing that the definiens does not reproduce that sense which, in use, is linked with the definiendum; for as a convention about the use of a sign, a definition cannot be false or true in the sense in which an empirical assertion is. We must therefore enquire whether the concept of language as ordinarily understood does not already carry with it certain more or less confused principles of grammar; principles, that is which do not arise from the fortuitous constellation of events (here, now and thus) but from the world's invariably being thus.

Even in this shape the question is not yet formulated precisely enough. Indeed, it leaves as yet undetermined what kind of 'origin' is meant here; for since from the outset we have stressed the signitive character of linguistic expressions and, with Husserl, we have distinguished between 'sign' and 'image', we know that there need be no similarity between linguistic signs and that which they signify, so that from the structure of the world we cannot simply infer features of the structure of a symbolism that is to express thought about the world. For example a single note may mean the existence or non-existence of a fact (a whistle signal). On the other hand there will have to be a certain similarity connections between symbol and symbolized, if a symbolism is to satisfy a certain condition (to be specified presently) and so to 'achieve' something. Whether calling the symbolism a 'language' is to hinge on the fulfilment of that condition depends on conventional decisions that are indifferent as regards the logic of language. The fundamental condition tacitly assumed as at the base of the idea of a pure grammar is the following:

We must fix principles of sentence formation[9] such that for any arbitrary word newly introduced the mode of its use within the framework of syntax is fixed in advance on the basis of its meaning, so that for any arbitrary object of thought linguistically fixed, its 'inner and outer' connection with other arbitrary linguistically fixed objects can be linguistically exhibited according to fixed rules. To satisfy this postulate, the syntactic rules must in some measure be adapted to the world's structure (foundational connection, spatio-temporal determinacy); for only the formal framework of the world is definitely surveyable, whereas possible factual variants of what is are unlimited.

Since determinacy of content includes determinacy of structure, in a syntax that takes account of all possible structural variations the syntactic use of any significant sign newly introduced is unambiguously fixed. Accordingly we distinguish nouns, adjectives and so on; as soon as the meaning of a word has fixed it as a noun, adjective and so on, we know its use in the formation of sentences.

Since in a symbolism of the kind just specified what constitutes the formal structure or constant frame of any extensions of language is syntax, to express that an object figures in a state of affairs we must introduce into the latter the word denoting the object.

For a certain multiplicity of the aspects making up a state of affairs, there is thus a minimum number of words or mutually independent parts of words that must figure in any sentence asserting this state of affairs while satisfying our condition of a fixed syntax, and to this extent there is a point in speaking of 'similarity' between the world and its representation in language. Nevertheless it is not appropriate to call language a 'logical picture of the world'.[10] Rather, in order to avoid conceptual errors in the logic of language, one must always keep in mind the signifying character of language and consider that there is no need as such for similarity between linguistic sign and what it denotes: such a similarity can show itself as necessary or appropriate only with respect to certain demands that, consciously or not, we make on language. In this we must once again distinguish between the reasons for this necessity or appropriateness, as to whether they are rooted in the changeable facts of experience or in the pure being so of the world. The latter reasons then constitute the *a priori* foundation of grammar. This formulation must therefore not be viewed as though the linguistic expression for the existence of facts (sentence formation) is fixed *a priori*; rather, it serves to emphasize that the syntactic rules concerned spring from ontological sources. The requirement specified above is in principle satisfied in all well-developed languages and so occasions no acute problems for the logic of language; without causing significant difficulties, one could make the requirement stricter still by adding the condition that even without knowing the full meaning of the individual linguistic signs we should be able to infer from them how to connect them with others in order that the result can be a sentence, that is, the linguistic expression for the existence (or non-existence) of a fact. For a requirement that is not fully realized in modern languages is satisfied if we simply mark the words that mean corporeal things, sensible qualities, positional determinations and the like, by certain prefixes or suffixes, and then tie the rules for sentence formation to these external characteristics. Of course, if, given the syntax, the semantic function of words is attached exclusively to structural aspects, then there is no need, for the enquiry in hand, to enlist their whole semantic character; but it would be a mistake to convey this insight by saying that it was enough to know whether a word was, say, a noun, adjective or adverb of place, in order to grasp its syntactic function, for the linguistic distinctions mentioned do not quite correspond to the structural divisions. To take a particularly important example, in most modern languages adjectives can be used as nouns; that is, syntactically like words that denote things. This discrepancy has decisive importance for problems in the logic of language, as is easily seen if alongside the postulate just analysed (that every state of affairs consisting of linguistically fixed elements is to be expressible by means of the syntax laid down once and for all) we place the complementary postulate that every sentence formed from words of the language according to the rules of that syntax must be meaningful (must say something about the world or expresses a result of reflection).

We see at once that this requirement is by no means satisfied by modern everyday languages. Sentences like 'the deltoid is virtuous', 'the colour "blue" is seven feet long', 'high *C* is as hard as diamond', are evidently senseless; 'for' – so the explanation usually runs – 'we cannot predicate moral qualities of geometrical figures, length of colour or hardness of sound'. However, this 'explanation' is apt to mislead us because it conceals the radical insight that 'about' geometrical figures, colours and sounds as such we can predicate nothing at all.

That this fundamental insight has not yet become common ground amongst philosophers is intelligible chiefly in view of the continuing lack of a proper grasp of the nature of generality, which in turn can be explained in terms of the intimate link between this problem and the central problem of philosophy, namely the question of clarification. The ability, essential for thought, of grasping what is general separately from its 'realization' in any given instance, of recognizing it afresh and varying it within certain limits, has led us to create a false co-ordination between individual and general objects, and to assign the same 'properties' and 'relations' to both. Nor could the distinction between 'external' and 'internal' properties and relations bring about radical change here, so long as the recognition of generality had not been clarified in principle; that this clarifying task has been carried out by Husserl[11] is therefore highly important for the problem of the foundations of logic and mathematics. We therefore recognize that if a language is to express the existence or non-existence of empirical facts by means of predicative sentences, and structural differences are to find expression in syntax alone, the only subjects that can figure in the language will be individual objects (individual concrete physical and psycho-physical entities). For simplicity we can assume that a word of the language is marked as a subject externally (for instance by its ending).

This prescription alone does not of course ensure that every concatenation of words obeying it will issue in a sentence; to achieve that, we should have to give several other easily formulated rules for the other parts of a sentence as well (predicate, object). However, failure to obey the prescription is a main source of the difficulties that have arisen in the foundation of mathematics through the discovery of circular definitions and contradictions. For both circular definitions (the use of impredicative concepts) and contradictions are tied to 'properties of properties', 'functions of functions', 'sets of sets' and the like, that is, to concept formations that in view of our earlier observations are inadmissible. More accurately: there is no connection between property and sub-property conformable with the connection between thing and property. This goes not only for properties in the narrow sense, but for any abstract entities: thus, to use the terminology of mathematical logic, one propositional function cannot be an argument of another.

An application of this insight that is particularly important for logic concerns 'judgment about judgments' and 'concepts of concepts'. A judgment in the logical sense arises by abstraction from the incidental features of the act of judging, such as the person judging, place and time of judging, and by setting aside the confusion pervading all thinking, while retaining only the adequation of assertion to what is, namely the truth of the judgment. However, this abstraction must not be re-interpreted as though there were, alongside what is, a separate domain of truth or realm of validities that links thought and being: judgments in the logical sense and truth as such do not stand above the psycho-physical subjects and things about which the judgments run, but are merely expressing invariance with regard to the different makers of judgments, the spatio-temporal location of acts of judgment, always on the assumption of perfect clarity of thought. Therefore to assert that a judgment is true (false) is an abbreviated formulation for saying that anyone anywhere anywhen making this judgment judges correctly (incorrectly).

Likewise for 'concepts of concepts'. A concept is the act of meaning an object with perfect clarity, abstracting from all occasional features of the intentional act. Hence we cannot say anything about this act of meaning that is not already contained 'in it'.

Here as at many other points of the problem area of the foundations of logic and mathematics much confusion has been wrought by merging a concept with its linguistic symbol, that is, with a word. This merging has, however, deeper reasons that we shall consider presently.

One might argue as follows: no logical sentence says anything about the world and no result of logical considerations can be touched as regards truth by the course of events; such sentences are therefore trivial in the strict sense of the word. The fact that ascertaining these trivialities is indispensable to thought, rests on the ability of language to obscure both connections and differences in sense. Logic viewed as a task (goal) is thus the creation of a language free from such mistakes, and logic as a system is the system of grammatical rules determining this language.[12]

However, this way of arguing already contains the seed of its own defeat. For it occasions immediately the question, what are the presuppositions for a language's being free from the mistakes just criticized? The determination of these presuppositions, which is the aim of logic, can be achieved only through clarification of the prior phenomenologico-ontological data. It is this layer of problems that supplies the peculiar theme of logical and mathematical theory. The logic of language thus provides no substitute for philosophy but quite on the contrary a particularly favourable way of access to it.

Having thus located the problems here concerning us, we shall survey some of the most important problems that have become acute in the foundational controversy, as far as they concern logic in the narrow sense, and show that they lose their sting as soon as we have gained insight about the relation of language to logic as above explained. We shall begin by analysing the sense (epistemic content) of the sentences of logic.

II. LOGICAL SENTENCES AND PRINCIPLES OF LOGIC: THEIR SENSE

The insight that logical sentences say nothing about the world raises the urgent question where we must look for their epistemic content. To call them tautologies, as has lately become the custom, expresses the view that their provenance rests on saying the same thing in different words and recognizing the thing thus differently expressed as the same. Take for example the logical sentence '*p* or *q* is *q* or *p*', where *p*, *q* are sentences and '*p*' is to mean '*p* is true'. The truth content of the sentence given could then be interpreted as follows: knowing what is meant by 'or', one recognizes that '*p* or *q*' denotes the same thing as '*q* or *p*'. Logical cognition resides in the assertion of this identity of meaning. However, this interpretation is untenable, for the meaning of '*p* or *q*' is not from the outset distinguished from that of '*q* or *p*', so that there is in the sphere of meaning nothing different that is subsequently equated. What is different is merely the symbolism. Since we can infer no meaning from the acoustic or visual 'stuff' of the symbols as such, neither can we infer sameness of meaning of acoustic or visual complexes from these complexes themselves.

From this it seems to follow that logical sentences are nothing but stipulations about the use of symbols, or consequences of such stipulations. This view is strongly supported by the brief account of the development of axiomatic theory in symbolic logic as given in the introduction. The logical calculus, interpreted in the sense of Hilbert's formalism, starts from figural elements and certain complexes formed from them (basic formulae), and contains prescriptions for forming further complexes (formulae) from them. In this we abstract entirely from the sign-like character of the formulae or their figural elements, but if 'afterwards' these elements are interpreted as signs for certain meanings fixed once and for all (not, or, and, implies, function, and so on), then the figure reached by the concatenation of formulae according to the prescriptions becomes a proof, and any 'formula generated' in the calculus becomes a formula provable from the basic formulae.

From this it seems to follow unambiguously that meaning is quite irrelevant as regards logical inference, for the inferential process was performed without use of meanings (mechanically), the latter being inserted in the results afterwards. Moreover the sense of the so-called logical constants 'not', 'and', 'or', 'implies' seems to be defined solely by the set of transformations of formulae that contain them.

Closer analysis, however, shows that this is not so. To see this clearly we must first remove a series of misunderstandings that may at times be linked with the calculus of logic as conceived by formalism. We must begin by ascertaining that the kind of figural elements is unimportant as regards the cognitive purpose that the calculus is to serve. It is obvious at once that one might replace the figural elements of the Russell–Hilbert calculus by other figural elements, without anything changing as regards logic, if, but only if, the number of different figural elements and the rules governing the formation of formulae remain unchanged. Thus we recognize that the symbolization of a definite multiplicity of different elements is essential for the calculus, but the visual quality of the individual symbolism is completely irrelevant.

On this point, those concerned with foundational enquiry are now generally clear, but almost wholly not so as regards the question about the cognitive content of the calculus. Here we must distinguish very sharply between the question as to the cognitive sources and that as to the thought process in which a certain item of knowledge is acquired. Mathematical proof as a way of exhibiting something is time-bound and its analysis into elementary steps as carried out in Hilbert's proof theory provides the complete schema of such a temporal articulation, but the logical connection of the final formula proved with the initial formulae and inference rules is quite independent of the fact that it has been thus exhibited. Of course, this connection is not a matter of having to view the initial formulae as premisses and the inference rules as syllogistic; for we have abstracted from any meaning of the formulae (which could contain some other meaning). Rather, the inference reads as follows: a 'provable formula' is defined as a figural complex that belongs to a certain transformation group as regards given figural complexes (basic formulae). F_1 belongs to such a transformation group = F_1 is a provable formula. The term 'provable formula' must here be taken as devoid of meaning; that a figural complex is a provable formula is to mean no more than that certain index F_p is assigned to it. The rules are now so chosen that on insertion of certain logical concepts instead of the figural elements the 'provable formulae' become logical sentences derivable from the basic logical sentences (into which the initial formulae are transformed by the insertion).

As against earlier mathematical logic, 'formalization' as carried through in Hilbert's proof theory lies in abstracting in its formulae from the meaning of logical concepts; by this move the total or partial identity of meaning between basic and derived sentences becomes a community not further defined which can be symbolized by a common index (F_p).

This, in principle, clarifies the sense of 'formalism', but this is of no great advantage for our efforts to grasp the sense of logical sentences, since, as we saw, in these formulae one abstracts from the logical sense of logical concepts. We must now attend to these formulae and enquire what is the nature of the concepts that appear in them, so that we may grasp the sense of logical sentences. Let us begin by considering the so-called logical constants 'not', 'and', 'or', 'implies', because in recent years they have been the object of investigations that have led to a result that seems definitely to solve the question as to their sense.

Wittgenstein's view on this matter is that logical constants say nothing about the world. It relies in particular on the following argument: the meaning of concepts is obtained from their use in sentences and the sense of these is just the method of their verification. The truth or falsehood of sentences in which logical constants occur is, however, easily seen to be unambiguously determined by the truth or falsehood of sentences in which no such constants occur. Sentences that do contain logical constants are thus 'truth functions' of sentences that do not. To begin with we recognize that the sentence not-*p* (*p* does not hold) is true when and only when *p* is false, and false when and only when *p* is true. For the sentences '*p* or *q*', '*p* and *q*', '*p* implies *q*'[2] the dependence of truth values from those of sentence *p* and sentence *q* is shown in the following schema, which is easily extended to any number of sentences.

pq	*p* or *q*	*p* and *q*	*p* implies *q* (not-*p* or *q*)
TT	*T*	*T*	*T*
TF	*T*	*F*	*F*
FT	*T*	*F*	*T*
FF	*F*	*F*	*T*

The distribution of truth values of the sentences *p* and *q* therefore unambiguously yields the truth values of the compound sentences that contain logical constants and no room is left for logical constants having a sense of their own.[3]

However, Wittgenstein's analysis is no definitive solution of the questions as to the sense of logical constants, for on deeper reflection the following consideration emerges: consider an empirical sentence containing a logical constant, say, the sentence 'in this room there are roses and carnations' (an abbreviation of the compound sentence 'in this room there are roses and in this room there are carnations'), what is the meaning of the logical constant in this sentence? As a first approximation one might say that the 'and' means a gathering together, but this concept needs to be clarified much further. What seems certain is that the 'and' relates in a definite way to the 'world', for it links concepts of things (roses, carnations) and it would be incomprehensible that a sentence as linguistic expression of an empirical fact should contain elements relating not to facts but to other sentences. However, in view of the undeniable cognitive fact (brought to light by Wittgenstein) that sentences that do contain logical constants are truth functions of ones that do not, we are driven to the conviction that thinking the world (language being merely the symbolic expression of this, and nothing more) involves certain principles that add nothing to the cognitive data but serve our grasping how they are connected, in ways presently to be specified.

We here have two pairs of opposite aspects: identity, difference; and universality, particularity. To begin with, take the first pair. If we consider two things and observe that they have the same colour, nothing is added to the contents of the perceptions of these objects, for it is precisely on them that the identity of colour depends; but the possibility of this observation is tied to conditions not yet contained in these perceptions as such, for the comparison on which the assertion of identity rests requires that we keep the perceived aspects in mind and isolate them in abstraction from other elements linked in the unity of perception, and therein doubtless lies a cognitive source of its own kind.[4]

It is now easy to show that the above mentioned logical constants express nothing but observations of identity or difference.

Consider first negation. Any 'genuine' (empirical) sentence constates that at a certain place there exists something of a certain kind. Without impairing the universal validity of our enquiries we can, for simplicity of formulation, confine ourselves to sentences about the external world and therefore assert that an empirical sentence contains the constatation that at a more or less sharply defined spatio-temporal location there exists something of a certain kind. The empirical constatation that amounts to the sense of negation is then that qualities of objects within a certain spatio-temporal domain are different from certain qualities linked with these objects in thought. This can be expressed either by the sentence 'amongst the objects at a given place and time there are none having the qualities mentioned' or by the sentence 'objects of the qualities mentioned are not at the given place and time'. Negation is thus the constatation of a difference.

In contrast, conjunction constates an identity, whether of the domain of individuation or of qualities. Thus the sentence 'objects O_1 and O_2 are yellow' means 'O_1 has the same colour as O_2, namely yellow' and the sentence '*A* and *B* are in this room' means '*A* and *B* are in the same space, namely in this room'.

If we are to understand the meaning of negation and conjunction in logical and mathematical thinking, we must keep in mind that here internal properties take the place of 'external' ones and that accordingly any negation means an inner difference which if ignored leads to absurdity (contradiction). We may further point out that the universal practice of linking premisses in syllogism or axioms in a calculus by 'and' is intended to express identity of 'logical function' within the framework of the inferential process. Since in the logical calculus it is shown beyond objection how the concepts 'or' and 'implies' can be expressed by 'not' and 'and', we can omit illustrating these observations in the case of disjunction and implication, and proceed at once to the pair of concepts universality, particularity.

That every determination that holds for the universal also holds for the particular is not a piece of knowledge, if we presuppose that the sense of the two terms is given; for that is precisely what marks this pair of correlative concepts. Nevertheless, this sentence expresses a fundamental insight, which lies in the implicit possibility of recognizing the universal. That a being-so (say, a certain colour) can be mentally severed from the real context in which it occurs and can be considered in isolation and varied in certain directions (hue, brightness, saturation), is a cognitive fact (in phenomenology called 'ideation' or 'seeing the essence'), that gives sense to the universal in the first place. This shows with special clarity that the sense of logical 'pseudo'-sentences, just like that of 'genuine' empirical ones, can be radically grasped only by going back to the cognitive sources. These do not, however, lie merely in the having of sense experience but also in the operations of spontaneous thought. To return to the pair of opposites: that the particular is 'contained' in the universal is a trivial observation; but the knowledge of what it is to be universal, which is presupposed in this observation, involves an insight of basic importance.

In this connection we must carefully consider that cognition of what is universal concerns not the universal as such but always some definite universal. Generalization stops with a concept that has empirical content (for example colour, sound) and what corresponds to the concept of the universal as such is no longer an object of thought but an operation of thought. Accordingly Husserl distinguishes sharply between generalization and formalization, or between determination and deformalization.

We can formulate this insight as to the sense of logical sentences as follows: logical sentences have no cognitive content, but the meaning of logical concepts consists in certain schemata for certain mental operations essential to cognition.

This likewise opens up an understanding of the sense of logical principles, that is, the laws of identity, contradiction and excluded middle. Here too, as we shall presently show, we do not have assertions whose truth could be described as cognitive, but rather the cognitive content of these sentences lies in their reference to fundamental mental operations.

Consider first identity: every object is identical with itself. The triviality 'behind' this insight is reached as soon as we examine the mental operations underlying the identification. It is only quite recently that this question has been clarified, by the analyses in Husserl's *Formal and Transcendental Logic*, which have uncovered the 'retentions' and 'protentions' that underlie the constitution of the object as 'identical with itself'. A reference to these profound and difficult enquiries must here suffice;[5] let us, however, insert a short comment on the role of the concept of identity in the standard work on symbolic logic, namely *Principia Mathematica*,[6] by Whitehead and Russell. Following Leibniz's identity of indiscernibles, they define identity as sameness of all properties, a definition with far-reaching consequences for the development of their system, since it is with the help of this concept that they define the concept of power set, which is essential for the ascent to the non-denumerable infinite and so for the whole of set theory. Wittgenstein[7] has given a stringent characterization of the mistakes in this notion of identity and in the second edition of *PM*, the authors point to the difficulties[8] that arise when Wittgenstein's criticism is taken into account. This criticism is that the identity of indiscernibles is not a logical principle and that identity is to be regarded neither as a property of nor as a relation between objects.

We now come to the analysis of the law of contradiction, which says that of two contradictory statements not both can be true: to grasp the underlying cognition here, we must focus on the concept of the contradictory. Of the two statements '*S* is *P*' and '*S* is not *P*' it is not absolutely necessary that one be false if for 'not *P*' we can insert anything that is not *P*. The two sentences '*O* (a definite object) is blue' and '*O* is pyramidal' may well both be true, but not the two sentences '*O* is blue' and '*O* is red' (both colour predicates taken to apply to the same part of the surface), nor the two sentences '*O* is pyramidal' and '*O* is spherical'. We see that the concept of the contradictory is here determined by that of the highest kind with objective content,[9] since contradictory predicates are ones that fall under the concept of that highest kind.

Another 'application' of the law of contradiction lies in the observation that no object can be in different places at once, or more precisely, not so unless it is also at all intervening places. Here too the 'applicability' rests on a definite not-*P*. Thus if the contradictory is correctly defined, the law of contradiction is trivial, but the mental operations underlying that concept give this 'triviality' an important sense.

Apart from this, the law of contradiction, as indeed logic in general, presupposes completely clear (adequate) thinking,[10] and it is this circumstance that in normative formulation lends to this law – as to the other logical principles – the character of a requirement. However, the possibility of this normative formulation is unimportant as regards the problem of the sense content of logic.[11]

From the phenomenological analysis of clear thinking it follows of necessity that it is absurd to envisage a contradictory (non-Aristotelian) logic whose relation to traditional logic would be roughly that of non-Euclidean to Euclidean geometry.

These observations include an unambiguous position as to the problem of the logical antinomies, for if logic as such cannot contain contradictions, their appearance must go back to a mistake of thought. Precisely in this problem area much confusion has arisen from a failure to distinguish sufficiently between logic and language (symbolism) and from consequently viewing apparent contradictions, generated by wrongly operating with a certain symbolism, as contradictions of logic. This seemingly rather imprecise observation becomes clearer as soon as we analyse the antinomies.

Following F. P. Ramsey[12] and A. Fraenkel,[13] logicians now divide these antinomies into two classes, the 'logical' and 'epistemological', the basis of classification residing in the need to use only logical (formal) concepts or other concepts as well when formulating the antinomies. We shall here consider only the 'logical' ones, which are by far the more important for the theory of logic. As paradigm for these interconnected antinomies we may look at Russell's so-called paradox[14] and use it to exemplify the following basic considerations. In this we do not base ourselves on a set-theoretical formulation (the antinomy of the set of all sets not members of themselves), but on the following, likewise given by Russell: Let a concept be called predicable if it can be asserted of itself, and impredicable if not. For example, the concept 'abstract' is predicable, the concept 'concrete' is impredicable. Let us therefore define predicable and impredicable as contradictory opposites so that any concept would have to be either predicable or impredicable. If now we examine the case of the concept 'impredicable' we find the following: if we assume that it is predicable, then the judgment ' "impredicable" is impredicable' holds, which states the opposite of the initial assumption. This must be wrong. Therefore it now appears as though it had been proved that 'impredicable' is an impredicable concept. Yet this judgment too is contradictory, for it implies that the concept 'impredicable' can be asserted of itself, so that it is predicable.

As is well known, Russell has eliminated this antinomy by his vicious circle principle, which is the basis of his simple theory of types. This principle states that no set can contain itself as an element; or, in terms of the more recent terminology of propositional functions, that no propositional function can contain itself as argument.[15]

It has been objected to Russell's theory that it erects a prohibition of thought, without giving reasons other than that it justifies itself in practice. (because by obeying it we avoid the so-called logical antinomies) without being able to show that the prohibition in its full scope is necessary for this. In fact the case stands quite otherwise.

To recognize this we must begin by remembering that talk about prohibition can relate only to rules of the logical calculus, namely operations with a certain symbolism, but not to logic itself, that is, to the operation of clarification described in the foregoing. Concentrating on the latter, we recognize – as follows from our earlier observations – that Russell's 'prohibition' is not only not wide enough but too narrow, for no propositional function can be argument of any other propositional function. In fact the failure to recognize this logical insight has had dire consequences in the Russell–Whitehead theory, especially when it constructs the theory of real numbers.

Against the thesis just put forward one might raise this obvious objection: if no propositional function can be an argument of any other, so that such a concept formation is senseless, how then is it possible that in certain cases, which Russell precisely wished to exclude by means of the simple theory of types, contradictions do arise, seeing that contradiction already presupposes that lowest level of sense whose principle Husserl calls apophantic logic? This objection is quite justified, since if the rules for forming meaningful sentences are thus disregarded for want of the presuppositions holding equally for contradiction and consistency, a contradiction can never emerge. However, the defective (ambiguous) use of symbolism, by feigning sense where there is none, can under certain circumstances produce a semblance of contradiction as much as of consistency. Let us make this clear by a critical analysis of Russell's paradox.

In this analysis we must take great care to avoid the following three mistakes of thought: (1) merging a concept with the objective aspect meant by it (object in the widest sense); (2) merging genuine determinations (external and internal properties) with mere names; (3) vaguely formulating the concept of 'contradictory opposition', for these formulations must be such (following our earlier observations) that senseless combinations or words are not confused with false sentences. Taking these points into account we obtain the following reformulation of the antinomies concerned: the concept C_a of an objective aspect a is called a predicable concept if 'C_a is a' is a true judgment. C_a is called an impredicable concept if 'C_a is a' is either a false judgment of if 'it' is a senseless combination of concepts.

This name-giving, which means no more than assigning an acoustic complex to the concept in question according to a certain rule, does not add any characteristic or new degree of determination to this concept, and indeed thought cannot alter its objects. Therefore we cannot assert its own names of an object of thought. Accordingly the concept 'good' by virtue of the nominal definition, is called impredicable, because ' "good" is good' is a senseless combination of words, but there is no question of its being impredicable. Rather, the sentence ' "good" is impredicable' is likewise a senseless combination of concepts.

Likewise with the concept 'impredicable'. ' "Impredicable" is impredicable' is a senseless combination of words and accordingly the concept 'impredicable' is called impredicable. It does not at all follow from this that the concept 'impredicable' is called predicable, for this would be so only if ' "impredicable" is impredicable' were a true sentence, which, as we just saw, it is not. The seeming contradiction thus arises from equivocation rooted in the ambiguity of the concept 'assertable', that is, in an inadmissible merging of thought and language, fact and arbitrary name-giving. In set theory the corresponding ambiguity is that of the concept 'set'.[16]

Analysis of the other 'logical' antinomies, most of which are closely related to the non-denumerable infinite of set theory, does not reveal any basically new mistakes of thought as regards Russell's antinomy and we therefore omit their detailed description and critical dissection.

With the dissolution of the antinomies, that is, the proof that the seeming contradictions go back to mistakes of thought, the properly philosophical part of this problem area has been settled. A quite different and much more difficult question is how to create a symbolism such that it permits a complete description of the object domain while making senseless combinations of concepts recognizable as such even externally.

This fundamental problem of symbolic technique in general and of logical calculus in particular need not concern us here.[17] We merely repeat our earlier observation that the main goal of our efforts will have to lie in the appropriate symbolic representation of 'internal properties' and 'internal relations'. A promising start for such a symbolism is now to hand in Wittgenstein.

One cannot emphasize too much that these problems belong to the technique of thought but not to epistemology and that extreme confusion arises if we do not sharply distinguish here.

We now come to the analysis of the law of excluded middle: of two contradictory sentences one is true. Since, like the law of contradiction, this rests on the concept of the contradictory, what was said there can be applied here. Here too the domain of factual variation (that is, the highest factual kind) plays an important role. Just as it is an 'application' of the law of contradiction that an object at a particular place and time cannot have two different colours, so it is an 'application' of the law of excluded middle that an object that at a particular place and time has not a certain colour C_1 must have another colour C_2. Just as it is an 'application' of the law of contradiction that the same object cannot be in two places at once, so it is an 'application' of the law of excluded middle that a real (existing) object that is not at a certain place at a certain time must at that time be at some other place. In this law we thus presuppose unambiguous qualitative determination and unambiguous spatio-temporal determination of what is, which means presupposing those mental operations that enable such unambiguous determinations to be made.

As already emphasized in the introduction, the universal validity of the law of excluded middle was challenged in the controversy concerning mathematical foundations. We shall now very briefly examine what this challenge amounts to.

If one views the two concepts 'provability of the statement *p*' and 'provability of the statement not-*p*' as contradictory opposites, the validity of the law of excluded middle within a certain formal system means that in that system either *p* or not-*p* is provable. On this formulation the law of law of excluded middle would for example no longer hold for geometry in Hilbert's axiomatic form, if of those axioms we omit the one corresponding to what is called Euclid's postulate (that through a point not on a straight line there is just one parallel to that line). As a result of this somewhat infelicitous terminology the validity of the law of excluded middle for a system thus coincides with the completeness of that system, that is, with the unambiguous formal definiteness of every basic concept contained in it. The question now is whether arithmetic (analysis) is a complete system in this sense, so that in it we can always prove either a relevant sentence or its denial.

E. Brouwer, whose name is associated with this 'crisis of excluded middle', denies this for existential sentences and distinguishes three cases: (1) the assertion that there is a number having certain properties can be proved by exhibiting such a number (construction); (2) it can be proved that the assertion that there is a number with these properties is absurd; (3) it can be proved that the assertion that the existential assertion is absurd is itself absurd, without it being possible to indicate a number satisfying the existential assertion.

The decisive question now is whether this third case is indeed possible, that is, whether (as Brouwer maintains) the inference from the absurdity of the absurdity of an existential sentence to the holding of that existential sentence (producibility of an example) is inadmissible.

I have elsewhere[18] tried to show that this trichotomy does not actually exist, but that, contrary appearance due to symbolic abbreviations notwithstanding, every proof of the absurdity of the absurdity of an existential sentence implicitly contains the construction of an example satisfying that existential sentence. If this argument should turn out to be valid, the central problem of foundational controversy would thus be solved. We cannot here go into detail on this but must confine ourselves to remarking that the far-reaching philosophic inferences that have been drawn from these problems are without exception untenable.[19] Above all it is absurd to say that the law of excluded middle (or any other sentence of logic) is not valid for a certain domain, for 'sentences of logic' are, as we have established, not sentences in any genuine sense; they are not vehicles for assertions that could prove themselves in some domains but not in others. What they actually are should be, in terms of the foregoing enquiry, as clear as we can make it at the level of these analyses. The ultimate clarifications are the task of transcendental phenomenology.

III. LOGIC AND MATHEMATICS

Turning now to the last part of our enquiry, namely an analysis of the relation between logic and mathematics, we must distinguish two main questions: (1) has mathematics specifically mathematical concepts, over and above those of logic? (2) is there a specific method of mathematics that is alien to logic?

These are questions as to how far mathematics in its concepts and methods goes beyond logic, for we can take it as uncontroversial that all concepts and methods of logic figure in mathematics. Since the reduction of the various geometries – that is, certain systems of formal relations to parts of arithmetic – no longer constitutes a serious problem, we shall confine enquiry to arithmetic (including analysis) and observe further that the whole of arithmetic and analysis can be reduced to the theory of natural numbers, an insight that has by now become common ground in foundational enquiry.

Thus the problem of the relation between logic and mathematics issues in the two main questions whether the natural numbers are logical concepts and whethei in mathematical operations with natural numbers there are extra-logical principles.

As to the first question, consider by what route of abstraction we reach the natural numbers. We recognize directly that these numbers are 'given' in the process of counting, which consists in one-one correlation of certain objects with certain distinguishing marks used in a prescribed sequence. We can therefore describe the counting process as follows:

Given various strictly distinct things (T), as well as other things different from these and from each other, which latter we shall call signs (S).

The following stipulations are to hold: to each T one and only one S is assigned as far as the supply of S's lasts. We stipulate which is the first S that is assigned and, for each S, which is the next S to be assigned; but there is no stipulation as to which T has which S assigned to it. (The mode of one-one correlation is arbitrary.) What is invariant to the various modes of such correlation is the S assigned to the last T present (or, if there are not enough S's, the last T entered into the correlation). Assume that there are enough S's; if T's to be denoted are unambiguously fixed, then this determines one and only one sign S that is the sign of the last T at each stage.[1]

Further enquiry now shows for a start that the kind of distinguishing sign is in principle indifferent, and next that 'external' signs are theoretically irrelevant; as regards epistemology, the only essential aspect is the fact of discrimination by signs, that is, the observation and retention of differences. Finally, reflection shows that the temporal aspect residing in the successive denoting of counted objects must be irrelevant for mathematics, since obviously no temporal aspect is involved in the propositions of mathematics. Whereas the first two observations introduce no essential difficulties as regards our earlier results concerning the relations between thought and language, the requisite exclusion of the concept of time seems to lead to very serious problems, since if numbers are not 'generated' in time, so that the model for enumerative order does not lie in temporal order, we are forced to assume that a definite number is a property belonging as such to a complex of things independently of the counting process.

Along this line of approach one is apt to define number as a set-like quality of a complex and in general to regard the concept of set as prior to that of number. This is the view expressed in Georg Cantor's set theory and in the logistic theory of Russell which is largely guided by it. However, this view is in principle untenable, quite apart from the fact that by merging individual and specific universality it has led to dire consequences: for obviously a complex of things itself already constitutes a collection and accordingly cannot have properties alongside its constituent elements. One must therefore ask oneself whether the choice between 'time-bound counting process' and 'set-like quality' or 'class of classes' is an unavoidable alternative in the definition of natural numbers,[2] or whether between the Scylla of radical subjectivism and the Charybdis of radical objectivism the right pathway of thought lies open.

Here, our earlier considerations concerning logical concepts (identity, difference; universality, particularity) take us forward by a decisive step, for our analysis has made us recognize that operating with the opposition 'subjective' and 'objective' is inadequate to clarify these concepts. For this presupposes on the one hand spontaneous mental operations that could be called subjective aspects, and on the other hand the 'matter' for these mental operations is formed exclusively by 'objective' (receptively apprehended) aspects.

One of these logical concepts doubtless does enter into the definition of natural numbers, namely that of difference. This we can infer directly from the analysis of the counting process immediately above. What remains to be examined is how far the concept of difference can take us in the analysis of natural numbers.

To ascertain this let us proceed as follows: let us start in thought from a certain thing A and observe of a thing B that it is different from A. We then call B 'a second relatively to A' and A 'the first belonging to B'. This initially somewhat strange nomenclature ceases to be so as soon as it is put alongside the pair of concepts 'the one, the other', which obviously expresses difference and nothing else and is distinguished in traditional usage from the pair 'the first, the second' only in that the latter contains a temporal direction of apprehension while the former does not.

Further, we call an object *C* a third relatively to a 'second' (*B*) and to the 'first' *A* belonging to *B*, on the basis of this observation: *C* is different from *B* and from *A* which is different from *B*. Quite in general the meaning of 'counting on' is the indication of an object quite different from the mutually different objects already given. If within the 'counting process' so defined an object *O* is the *n*th relatively to certain other objects we shall say that we have counted a 'complex of at least *n* objects'. If further in this case none of the counted objects is different from *O* and from the objects that go with it, we shall say that we have counted a 'complex of exactly *n* objects'. We now recognize at once that the above definition of the 'counting process' is an adequate description of counting, for the 'signs' that were important in our introductory analysis are, as observed earlier, merely a means of distinguishing but otherwise in principle inessential. The thinking process that I perform when I enter a thing into an enumeration, is in fact no more than the observation that the thing is different from the others (those already counted).

We can now further understand that in a counting process in which it has been fixed which objects are being counted (that is, in which for every object it is decided whether it shall be counted or not) the sign (ordinal number) of the last element remains the same, in whatever order the objects are counted; for this follows directly from the symmetry of the difference relation; that is, the fact that '*A* is different from *B*' means the same as '*B* is different from *A*'. It is such 'conversions of difference relations' that constitute the permutations of the counting process which are here involved; for that the aspect of temporal variation as such is here irrelevant can be seen by formulating the invariance in question as follows: if the objects to be counted have been fixed, the signs used are the same whichever of them is assigned to whichever object in one-one correlation. That this reformulation is admissible becomes clear if we consider that the re-ordering in the counting process manifests itself merely in that a different sign is now assigned to any definite object. Therefore we have definitively determined the basis on which we recognize this invariance, which some mathematicians have tried to prove while others (as for instance Brouwer) have regarded it as an immediately obvious cognitive principle.

Thus the concept of any definite natural number is fixed by a definition of use in Russell's sense, and the 'properties' of natural numbers on which we base an axiom system of arithmetic in Peano's sense follow directly from the foregoing observations.

For from them it follows that: (1) there is just one first number; (2) between any number and its immediate predecessor there is a one-one relation; (3) the structure of the number series (the form of the natural numbers) is determined exclusively by stipulations (1) and (2).

Comparing this with the classical axioms of Peano[3] (which contain the three basic concepts '0', number', 'successor'): (i) 0 is a number; (ii) the successor of any number is a number; (iii) no two numbers have the same successor; (iv) 0 is not the successor of any number; (v) any property of 0 belonging to the successor of any number having it belongs to all numbers; we easily see that (1) corresponds to the conjunction of (i) and (iv) and (2) to the conjunction of (ii) and (iii).

What is more difficult is to see that (3) corresponds to (v), the so-called principle of complete induction, but I have elsewhere[4] shown that this is so and must here refer to that demonstration.

In considering the principle of complete induction we have reached the question whether there is a specifically mathematical method; it is above all this principle that is alleged to contain the collection of an infinite number of syllogisms (Poincaré) and therefore to allow inferences that escape finite logic or to be regarded as a specifically mathematical source of knowledge. On that view, in operating with universal propositions and existential propositions in mathematics we bring this transfinite principle into action.

In rebutting this misconception we must be very brief.[5] It is essential to distinguish between universal and existential propositions here:

First, universal propositions: these refer not to an infinite totality of numbers but to the general form of number, that is, to a formal schema. The mistake in thought involved is a special case of the merging of individual and specific universality. However, the problem is more complicated still. To see this we must briefly digress on the completeness of mathematics.

From the general form of natural numbers as described by Peano's axioms suitably interpreted, it follows in particular that the system of natural numbers is complete, that is, any natural number whatever is structurally fixed without ambiguity, and this is seen on the basis of the following consideration: the determinations that there is just one first element and that the relation between immediate precursor and immediate successor is one-one, leave no further possibility of variation; and since further determinations are excluded, this means that no determination leaves open the possibility of variation and that only one single formal structure satisfies the stated conditions. This is precisely what completeness amounts to, which on the basis of seemingly different definitions is variously denoted as monomorphism, non-branchability or definiteness of decision.[6]

That an axiomatic system is complete thus means that it is definite as to decision, that is, that any relevant proposition is either a consequence or non-consequence of the axioms. Therefore it becomes possible (against Brouwer's view) to 'apply' the law of excluded middle in arithmetic without limitation, so that from the absurdity of a universal proposition we can infer the existence of a counter-example.[7] Since on the other hand such a proof of absurdity must implicitly contain the construction of a counter-example, Brouwer's condition of constructivity throughout is satisfied in mathematics.[8]

This operating with the form of natural numbers and in particular the drawing of inferences from the completeness of the system contain aspects of thought in which people imagined they saw what was specific to mathematical method. However, there is no difference in principle with regard to logic properly speaking, which might justify an epistemological gap. For the concepts with which we operate in mathematics can be derived from the basic concepts of logic (identity, difference; universality, particularity) and the 'method', as in logic, is one of clarification of propositions. Nor is this refuted by pointing to the fact that mathematical enquiry has, for the purpose of clarification, created miracles of cogitative technique, in particular a symbolism that is in practice almost indispensable. For even the most complicated proof is a chain of logical inferences and all symbols different from those for basic concepts are defined in terms of those and therefore eliminable. In spite of all this, it looks as though a further sound argument might be advanced for the essential difference of mathematics from logic, namely the argument of 'potential infinity' in the number series. This is the proposition that for any number there is a greater number. For logic can in no way contain an analogous proposition, since the latter is not a tautology.

This brings us to existential propositions, the crux of foundational enquiry in mathematics, but in view of what has already been said a few very brief remarks will suffice. For according to the thesis (put forward in this essay and argued in detail elsewhere) that every proof of an existential proposition contains the exhibiting of something, the existential concept as basic to mathematics simply disappears. A proposition of the form 'there is a number with the "property" P' thus means, when viewed as the starting point of a problem, 'let us look for a number with the property P'; and as the solution of a problem, 'a certain indicated number N has the property P'. Thus the proposition 'for every number N_1 there is a greater number N_m' has no other sense than the exhibiting of such a greater number, and this is just the insertion of it into the general schema of number and therefore presupposes no synthetic *a priori* judgment.[9] It thus remains the case that mathematics is a vast tautology.[10] As pointed out earlier, the so-called extensions of the number domain are merely symbolic abbreviations. All propositions concerning 'new' numbers therefore can be translated back into propositions in which only natural numbers occur.

The only one amongst these extensions that continues to cause significant difficulties to foundational enquiry is that of irrational numbers. As I have shown elsewhere,[11] these difficulties are connected with the defective class theory of numbers and vanish in principle when that has been eliminated, even if this or that individual case is not readily clarified because of the intricate symbolism. For the source of these difficulties lies in the iteration of the concept of class (set, property, function) and therein lies the common origin of the antinomies, existential propositions and impredicative (non-predicative) concept formations.

On the last of these, let us add a few basic comments by way of conclusion. By a 'non-predicable concept formation' we understand "quite generally the formation of two concepts in such a way that the definition of either must contain the other."[12] Setting aside the sphere of 'concepts' inserted between thought and language and objects thought or mentioned respectively, this is readily seen to mean that the object intended by a sign (pseudo-sign) S_1 is to be entirely or partially determined by indicating an object intended by a sign S_2 which object in turn is to be determined by the object denoted by S_1; which is obviously circular, and circular 'determinations' as such determine nothing.

However, the mistaken view that impredicative concepts could under certain circumstances be a usable means to knowledge is to be traced back to the fact that concepts have been regarded as impredicative when they actually are not. Of this, two examples. The first (following Fraenkel, op. cit. p. 249) runs as follows: "let $f(x)$ be a continuous real function of a variable for $0 \leq x \leq 1$, M the totality of all values of $f(x)$ belonging to the interval, m the least of these, that is the smallest number of the set of numbers M. (That in M there exists a smallest number, or, what comes to the same, that M is a closed set, is of course not self-evident but must be proved.)"

The second example comes from a recently published article by P. Bernays[12] and states: "In the totality of numbers, let a certain number be defined by the property that it is the greatest prime that multiplied by 1000 exceeds the previous prime multiplied by 1001."

We easily see that neither of these two typical cases contains an impredicative concept formation.

The least value a of the function mentioned is to be defined as follows: 'a is a value of the function f', 'any arbitrary value of f different from a exceeds a'.

The semblance of impredicativity arises by falsely identifying the function with the sequence of its values, instead of grasping that it is a mathematical form. This is in principle the same mistake as that underlying the sensualist theory of abstraction, whose continued effect beyond Mill as far as Russell has had a disastrous influence on the theory of mathematics.

The second case is to be solved in analogous manner. Our prime number p is to be defined as follows: 'p_m is a prime number whose thousandfold multiple exceeds the thousand-and-one-fold multiple of the previous prime number.' 'If p_n is any prime number greater than p_m, then its thousandfold multiple does not exceed the thousand-and-one-fold multiple of the prime number previous to it.'

Here the semblance of impredicativity vanishes as soon as we see clearly that 'prime number p_m' is a mathematical form that is not presupposed in the definition of p_m.

Bernays comments on this example "there is no objection as such to defining an object from a totality by means of a property that refers to that totality." "However, this presupposes that the totality in question is determined independently of the definitions referring to it; otherwise we land in a circular mistake." This presupposition is, however, unfulfillable in principle; no mathematical concept can be determined in such a way that this determination is independent of a definition referring to it, for every determination of a mathematical concept is an essential determination.

As we have seen, however, in the cases that Bernays considers there is actually no such 'reference', and that alone is the reason why we do not here land in erroneous circularity.

It can be shown that the concept of the upper bound of a set of real numbers (which is of basic importance in analysis) is actually defined without circularity, and that quite generally in analysis no circular inferences arise, provided we do not go beyond the domain of the denumerably infinite.[13] That the pseudo-concepts of non-denumerable sets have still not been completely seen through as such is understandable if we note that obviously legitimate concepts have been erroneously regarded as impredicative, which has led to the belief that impredicative concept formations, without which the 'ascent to the non-denumerably infinite' is impossible, cannot be simply dismissed.

In this way a theoretical analysis as regards the seeming circularity of logic and mathematics yields the same result as in relation to the antinomies; here too the semblance arises from mistakes of thought and vanishes when they have been dissolved.

NOTES

INTRODUCTION

[1] Thus in Hilbert's axioms of geometry, the basic concepts 'point', 'straight line', 'plane' are introduced as follows: we imagine three different systems of things: those of the *first* system we call *points*, denoting them by *A*, *B*, *C*. . .; those of the *second*, straight lines, denoting them by *a*, *b*, *c*. . .; those of the *third*, planes, denoting them by α, β, γ. . . .

Imagine points, straight lines and planes in mutual relations, denoting these by words like 'lying', 'between', 'parallel', 'congruent', 'continuous'; the strict and complete description of these relations is achieved through the 'axioms of geometry'. *The Foundations of Geometry* (tr. E. J. Townsend, II/1 Chicago 1921).

SECTION 1

[1] *Logische Untersuchungen*, 2nd ed. 1929, II/2, p. 55 [E.T. p. 711f.].
[2] Kant's principle of division which rests on a view of mathematics that is now known to be particularly wrong, could not be considered here; but Husserl's (*Ideen zu einer reinen Phänomenologie und phänomenologischen Philosophie*, Halle 1913), which rests on a distinction between formal and material ontology, could.
[3] Cf. especially II/1, p. 266ff. [E.T. p. 467ff.].
[4] Cf. Husserl's third investigation, l.c.
[5] l.c., p. 110f. [E.T. p. 340f.].
[6] Cf. below p. 221f.
[7] Cf. below p. 209.
[8] 3.325ff.
[9] We here confine ourselves to this part of grammar, which is by far the most important for foundational problems; thus we analyse the idea of a pure syntax.
[10] This expression is due to L. Wittgenstein, whose *Tractatus Logico-Philosophicus* is undoubtedly the most important contemporary contribution to the theory of logic and

mathematics. His analyses, growing out of the modes of thought of logicism, radically overcome the basic mistake of Russell's theory, which lies in the ambiguity of the concept of class, and contain important beginnings for a symbolism that takes this insight into account. However, as our account shows, his view as to the relation of language to the world is open to objection, witness the following extract:

4.12 Propositions can represent the whole of reality, but they cannot represent what they must have in common with reality in order to be able to represent it – logical form.

In order to be able to represent logical form, we should have to be able to station ourselves with propositions somewhere outside logic, that is to say outside the world.

4.121 Propositions cannot represent logical form: it is mirrored in them. What finds its reflection in language, language cannot represent. What expresses itself in language, we cannot express by means of language. Propositions show the logical form of reality.

They display it.

4.1211 Thus one proposition 'fa' shows that the object a occurs in its sense, two propositions 'fa' and 'ga' show that the same object is mentioned in both of them.

If two propositions contradict one another, then their structure shows it; the same is true if one of them follows from the other. And so on.

4.1212 What can be shown, cannot be said."

The mistake in this view (though of no great harm in connection with Wittgenstein's enquiries) consists in the misappraisal of the purely sign-like character of language. Language can no more 'represent' reality than logical form; but in view of the heuristic role that language (or a calculus) plays in the pursuit of certain goals of thought, it is expedient to symbolize logical form (structure) in a different way from the content given. The linguistic reasons for this expediency lie in the structure of the world, which is intuited *a priori*; but from this it by no means follows that a symbolism taking this insight into account is *a priori* correct.

[11] Above all in the second of his *Logische Untersuchungen*.

[12] For the identification of logic and grammar (syntax) the following train of thought is also important: since thinking is merely subjective and language objective, the objective validity of logic cannot be gathered from thinking only, but must be connected with language. The mistake in this argument lies in the ambiguous use of the concept of 'objective'. The 'objectivity' of language is rooted in the empirical fact that a more or less sharply defined domain of persons share a symbolism, the objectivity of logic resides in its being independent from all empirical data.

SECTION II

[1] In this connection a few words should be said about the relation of 'logic' and 'metalogic', or 'mathematics' and 'metamathematics', these concepts being taken in the sense of Hilbert's formalism. The contrast is usually effected by saying that logic or mathematics are described as a calculus and metalogic or metamathematics as the set of

statements about that calculus. However, from our analysis concerning 'statements about', it follows that these self-styled statements about the calculus can be nothing else than clarifications of the rules, that is, indications of what we are to understand by 'provable formulae'. Since, however, the application of rules, that is, the formation of 'provable formulae', likewise presupposes a clear grasp of the rules applied, there is no such difference of principle between logic (mathematics) and metalogic (metamathematics). The possible mechanical construction of probable formulae must not mislead us, for the 'mechanical construction of provable formulae' is nothing more than the mechanical construction of certain figures. That these figures conform to the rules can be observed only by clarification of the rules. Still, it is expedient to distinguish those clarifications that ascertain that a certain figural complex is a provable formula from the remaining clarifications (especially those that when interpreted concern the consistency and completeness of the system).

[2] This proposition, in the terminology of symbolic logic, means not-p or q, which is always true when p is false.

[3] From these considerations Wittgenstein derives his definition of tautology as a compound proposition that is true for any distribution of truth values of its elements, and of contradiction which is always false; as well as his theory of elementary propositions.

This states that all statements in the world can be compounded from simplest (elementary) propositions, which leads to important consequences especially as regards a symbolism that is to exhibit the structure of facts with clarity. We cannot here enter into closer discussion of Wittgenstein's most important enquiries relevant to this matter. For those who know his theories let me merely add the following remark:

(i) this doctrine seems to me to need completing as regards (a) the connection of universal propositions with the corresponding particular ones; and (b) the 'congruence of experience' underlying empirical statements, where the kind of elementary experience is mostly not unambiguously determined.

(ii) the assertion that one elementary proposition cannot contradict any other (4.211) seems to me to be false.

[4] The close connection with Kant's distinction between spontaneous and receptive acts cannot be discussed in detail here.

[5] *Formale und transzendentale Logik*, Halle 1929, p. 163ff. [E.T. p. 185ff.].

[6] Russell, B. and Whitehead, A. N., *Principia Mathematica*, Cambridge, vol. I, 1910 (new edition 1925), vol. II, 1912 (re-issued 1927), vol. III, 1913 (re-issued 1927).

[7] l.c. 5.4733 and 5.33–5.5352.

[8] Preface, p. XIV. Cf. Russell's introduction to Wittgenstein's *Tractatus*.

[9] Cf. Husserl, *Ideen*, p. 25 [E.T. p. 71].

[10] About the deeper problems that arise in analysing the 'idealising presuppositions' underlying the principles of logic, cf. Husserl, *Formale und transzendentale Logik*, p. 162ff. [E.T. p. 184ff.].

[11] Cf. *Logische Untersuchungen*, I, p. 9ff. [E.T. p. 58ff.].

[12] 'The Foundations of Mathematics', *Proc. Lon. Math. Soc.* **25** (1927), 338–384.

[13] *Einleitung in die Mengenlehre*, 3rd edition, Berlin 1928.

[14] First mentioned by Frege in *Grundgesetze der Arithmetik*, 2nd vol., Jena 1903, postscript p. 253 [E.T. p. 234]. For a detailed account see Russell, B., *The Principles of Mathematics*, Cambridge 1903, p. 101ff.

For a short summary of the logical and epistemological antinomies, see *Principia Mathematica*, vol. I, p. 63f.

Of the vast literature on the problem of the autinomies (detailed bibliography in Fraenkel, A., op. cit. p. 394ff.) we shall here mention only two especially important items: König, I., *Neue Grundlagen der Logik, Mathematik und Mengenlehre*, Leipzig 1914; and Behmann, H., 'Zu den Widersprüchen der Logik und Mengenlehre', *Jahresber. d. deutsch. Mathem. Ver.* **40** (1931), 37–48.

[15] A propositional function "is a logistic representation of a concept (as something that can be stated of an object). And propositional functions with one argument represent properties, those with two or more arguments two-term or multiple-term relations.

"Example . . . 'The Matterhorn is a mountain in Europe' is a statement, therefore 'is a mountain in Europe' or 'x is a mountain in Europe' is a propositional function with one argument, a property." Carnap, R., *Abriss der Logistik*, Vienna 1929, p. 4.

[16] Cf. Kaufmann, F., *Das Unendliche in der Mathematik und seine Ausschaltung*, Vienna 1930, p. 96ff., p. 190ff.; [E.T. this volume p. 80ff., p. 158ff.]

[17] On this point we refer once more to the paper by Behmann mentioned in note 14 above.

[18] Kaufmann, F., op. cit., p. 66f.; this volume p. 57f.

[19] Cf. *ibid.*, p. 64ff.; this volume p. 56ff.

SECTION III

[1] This cognitive fact was pointed out by Schröder, E., *Lehrbuch der Arithmetik und Algebra*, vol. 1, Leipzig 1873. Cf. also Stolz, O., *Vorlesungen über allgemeine Arithmetik*, 1885, part I, p. 9f.; Kronecker, L., 'Über den Zahlbegriff', *Werke*, 1889, III 1, p. 249ff.; Helmholtz, H., 'Zählen und Messen', *Wissenschaftliche Abhandlungen*, III p. 356ff.; Hölder, O., *Die Arithmetik in strenger Begründung*, 2nd edition, Berlin 1929, p. 14ff.

[2] The account in this section so far is a rather shortened version of the train of thought developed in my book *Das Unendliche in der Mathematik*, p. 77f.; this volume p. 68f. What follows is a deeper assessment of the results obtained there.

[3] *Arithmetices principia novo methodo exposita*, Turin 1889.

[4] Kaufmann, F., op. cit. p. 88 ff.; this volume p. 75ff.

[5] For this and what follows, cf. *ibid.*, pp. 90–105; this volume pp. 76–87.

[6] A similar consideration shows that Peano's axioms are consistent. It is obvious that axioms (i)–(iv) are consistent in pairs, since they stipulate that amongst natural numbers there is just one first number and that between any number and its immediate successor there is a one-one relation. Of these three determinations (the one-one relation contains two such) each determines something different from the other two, so that none can contradict the others. As far as I know, nobody has ever seriously considered that there might be contradictions within the first four axioms. Suspicion was always directed at the fifth axiom. As soon as we grasp that its sense lies in the exclusion of any further determinations, however, this mistrust has to vanish. For in the first place this condition cannot contradict any of the remaining axioms (since it does not refer to their content), and secondly the exclusion of any further determinations makes the consistency of axioms (i)-(iv) the consistency of the structure determined by them alone. The foregoing results as to consistency and completeness (definiteness of decision) of arithmetic seem to me entirely conclusive. I therefore cannot admit the demand that such analyses should be attached to a definite calculus and be carried out by its machinery.

On the other hand an explicit warning is called for against mistaking the proof of completeness (definiteness of decision) of arithmetic for the solution of the decision problem. That solution would mean "that one knows a procedure that for a given logical expression allows the decision as to universal validity or fulfillability through finitely many operations" (Hilbert-Ackermann, *Grundzüge der theoretischen Logik*, p. 73 [not in E.T.]). Particularly important on that point is a recent paper by Gödel, K., 'Über formal unentscheidbare Sätze der Principia Mathematica und verwandter Systeme I', *Monatshefte f. Math. und Phys.* **38**, fasc. 1, pp. 173–198.

[7] Brouwer himself calls it a central point of his doctrine that completeness is co-extensive with the universal validity of the law of excluded middle.

[8] This common abbreviated way of talking signifies that Brouwer reaches the reflective result that the assumption of pure existential propositions in mathematics is a nonsense.

[9] Of course there remain very deep problems in the background, as regards the idealizations of 'and so on' along with its subjective correlate 'one can always go on' (cf. Husserl, *Formale und transzendentale Logik*, p. 167). However, these problems of transcendental logic need not be taken into account in the treatment of questions as to mathematical method.

[10] Wittgenstein distinguishes between logic as a set of tautologies and mathematics as a logical method by which equations are transformed into other equations. The sense of an equation, according to Wittgenstein, lies in that it points to sameness of meaning of two complexes of signs, which must of course be recognizable from the complexes themselves. The sound core of these theses seems to me to lie in the distinction between logical constants and mathematical operators (+, −, and so on). For, contrary to appearances, these operators relate not to the numbers themselves, but to our thinking these numbers. On the other hand it seems to me that the character of mathematical cognition is not different from that of logical cognition, since in both we have the same operation of clarification (logical analysis).

[11] Kaufmann, F., op. cit., p. 119ff.; this volume p. 99ff.

[12] Bernays, P., 'Die Philosophie der Mathematik und die Hilbert'sche Beweistheorie', *Blätter für deutsche Philosophie* **4** (1930), 326–367, p. 356.

[13] Cf. Weyl, H., 'Über die neue Grundlagenkrise der Mathematik', *Math. Zeitschr.* **10** (1921), 39–79. "A real number is a set of rationals that correspond to a certain property of rational numbers. To a set of real numbers thus corresponds a property A of properties of rational numbers. The upper bound of this set of real numbers is itself the set of those rational numbers that possess a certain property R_a, namely the following: that there exists a property of the kind A which belongs to the number x". See also Kaufmann, F., op. cit., p. 130f.; this volume p. 107f.

BIBLIOGRAPHY OF THE PUBLISHED WRITINGS OF FELIX KAUFMANN

Compiled by Dr. Harry P. Reeder

This bibliography is based in part upon bibliographies provided by Prof. Lester Embree and Dr. Else Kaufmann.

Logik und Rechtswissenschaft, Tübingen, 1921, pp. xi, 134 (reprinted Aalen, 1961).

"Die theoretische Philosophie als Wissenschaftslehre", *Prager juristische Zeitschrift*, 2d. Jahrg., Heft 3/4, (1922), 124–8.
Recht (hereafter *ZöR*) **3** (1922–3) 236–263.

"Theorie der Rechtserfahrung oder reine Rechtslehre?", *Zeitschrift für öffentliches Recht* (hereafter *ZöR*) **3** (1922–3), 236–263.

[review of L. Nelson's *System der philosophischen Rechtslehre*] *ZöR* **3** (1922–3), 498.

"Die ökonomischen Grundbegriffe: Eine Studie über die Theorie der Wirtschaftswissenschaft", *Zeitschr. f. Volkswirtschaft u. Sozialpolitik* **3** (1923), 31–47.

Die Kriterien des Rechts, Tübingen, 1924, pp. iv, 161.

"Kant und die reine Rechtslehre", *Kantstudien* **29** (1924), 233–242.

[review of E. Landmann's *Die Transzendenz des Erkennens*] *Österr. Rundschau* **20** (1924), 33–39.

"Logik und Wirtschaftswissenschaft", *Archiv für Sozialwissenschaft und Sozialpolitik* (hereafter *AfSS*) **54** (1925), 614–565.

[review of C. A. Emge's *Vorschule der Rechtsphilosophie*] *AfSS* **56** (1926), 817–88.

[review of M. Salomon's *Grundlegung zur Rechtsphilosophie*] *AfSS* **56** (1926), 818–89.

"Staatslehre als theoretische Wissenschaft", *Kant-Studien* **31** (1926), 53–60.

Die philosophischen Grundprobleme der Lehre von der Strafrechtsschuld, Leipzig and Vienna, 1929, pp. viii, 138.

"Sociale Kollektiva", *Zeitschrift für Nationalökonomie* (hereafter *ZfN*) **1** (1929–30), 294–308.

[review of S. Landshut's *Kritik der Soziologie*] *ZfN* **1** (1929–30), 796.

Das Unendliche in der Mathematik und seine Ausschaltung, Leipzig and Vienna, 1930 pp. x, 203 (reprinted Darmstadt, 1968) [E.T. in the present volume].

[review of R. Carnap's *Der logische Aufbau der Welt*] *Archiv für Rechtsphilosophie* (hereafter *AfR*) **23** (1930), 200–202.

[review of R. Carnap's *Scheinprobleme in der Philosophie*] *AfR* **23** (1930), 202.

"Schuld und Strafzweck", [reply to Dr. Leopold Zimmerl's review of F. K., *Die philosophischen Grundprobleme der Lehre von der Strafrechtsschuld*], *Juristische Blätter* (1930) no. 6, 120–121.

[review of W. Burkamp's *Begriff und Beziehung: Studien zur Grundlegung des Logik*] *Logos* **19** (1930), 411–412.

"Note on Dr. Walter Schiff's 70th Birthday", *Neues Wiener Abendblatt* no. 149, 30 May (1930), 3–4.

[review of J. S. Mill's *Die Freiheit*] *ZfN* **2** (1930–1), 659.

"Was kann die mathematische Methode in der Nationalökonomie leisten?", *ZfN* **2** (1930–1), 754–779.

"Juristischer und soziologischer Rechtsbegriff", *Gesellschaft, Staat und Recht*, ed. Alfred Verdross (*Festschrift* for Hans Kelsen's 50th birthday), Vienna 1931, pp. 14–41.

"Bemerkungen zum Grundlagenstreit in Logik und Mathematik", *Erkenntnis* **2** (1931), 262–290 [E.T. in the present volume].

[review of Alfred Schütz's *Der sinnhafte Aufbau der sozialen Welt*] *Deutsche Literaturzeitung* **36** (4 Sept., 1932), 1711–1716.

[review of Benedetto Croce's *Logik als Wissenschaft vom reinen Begriff*] *Archiv für Rechts- und Wirtschaftsphilosophie* **26** (1932–33), 380.

"On the Subject-Matter and Method of Economic Science", *Economica* **13** (1933), 381–401.

[review of E. Husserl's *Formale und transzendentale Logik*] *Göttingische gelehrte Anzeigen* (1933) No. 11/12, 432–448.

[review of Albert A. Ehrenzweig's *Irrtum und Rechtswidrigkeit*] *Juristische Blätter* (1933) No. 5711, 1420.

"The Concept of Law in Economic Science", *Review of Economic Studies* **1** (1934), 102–109.

"Zur Methodologie der Sozialwissenschaften", *ZfN* **5** (1934), 241–245.

[review of E. Husserl's *Méditations Cartésiennes*] *ZfN* **5** (1934), 428–430.

Methodenlehre der Sozialwissenschaften, Vienna, 1936, pp. iv, 331.

"Remarks on Methodology of the Social Sciences", *Sociological Review* **28** (1936), 64–84.

[review of Emil Utitz's *Die Sendung der Philosophie in unserer Zeit*] *Internationale Zeitschrift für Theorie des Rechts* (hereafter *IZTR*) **10** (1936), 222–223.

[review of A. Fey's *Der homo oeconomicus*] *ZfN* **7** (1936), 559.

"Die Bedeutung der logischen Analyse für die Sozialwissenschaften", *Actes du 8éme Congrès International de Philosophie*, Prague, 1934 (published 1936), pp. 209–216.

"Die Phänomenologie der Kunst als Organ der Metaphysik", *Actes du Deuxième Congrès International d'Esthétique et de Science de l'Art*. vol. **1** (1937), 63–67.

"Do Synthetic Propositions *a priori* exist in Economics?", *Economica* **NS4** (1937), 337–342.

[review of H. Fried's *Über die Rechtsnot und ihre Bekämpfung*] *IZTR* **11** (1937), 56–57.

[review of F. Waismann's *Einführung in das mathematische Denken*] *Neue Freie Presse* 7 Feb. (1937), 22.

[review of R. Carnap's *Logische Syntax der Sprache*] *Sociological Review* (London) **29** (1937) 203–207.

"Über den Begriff des Formalen in Logik und Mathematik", *Travaux du 9éme Congrés International de Philosophie*, Paris, 1937, vol. **6**, pp. 128–135.

"The Significance of Methodology for the Social Sciences", *Social Research* (hereafter *SR*) **5** (1938), 442–463.

[review of Alfred Tarski's *Einführung in die mathematische Logik und in die Methodologie der Mathematik*] *Neues Wiener Tagblatt*, 30 Jan. (1938), 24.

"Unified Science (Note)", *SR* **6** (1939), 433–437.

"The Significance of Methodology for the Social Sciences (Part II)", *SR* **6** (1939), 537–555.

"Truth and Logic", *Erkenntnis* **9** (1939), 105–110.

"Phenomenology and Logical Empiricism", *Philosophical Essays in Memory of Edmund Husserl*, ed. M. Farber, Cambridge Mass., 1940, pp. 124–142 [E.T. by Dorion Cairns].

"On Dewey's Logic", *SR* **7** (1940), 243–246.

"Truth and Logic" *Philosophy and Phenomenological Research* (hereafter *PPR*) **1** (1940), 59–69.

"The Structure of Science", Journal of Philosophy (hereafter *JP*) **38** (1941), 281–292.

"Strata of Experience", *PPR* **1** (1940–1), 313–324.

[review of *International Encyclopedia of Unified Science*, volume 2, nos. 4 and 5 (Dewey and Woodger)] *SR* **8** (1941), 254–255.

[review of E. Husserl's *Erfahrung und Urteil*] *SR* **8** (1941), 256–258.

[review of S. Hook's *John Dewey*] *SR* **8** (1941), 393.

"The Logical Rules of Scientific Procedure", *PPR* **2** (1941–2), 457–471.

"On the Postulates of Economic Theory", *SR* **9** (1942), 379–395.

[review of P. Frank's *Between Physics and Philosophy*] *SR* **9** (1942), 542.

[review of J. Mayer's *Social Science Principles in the Light of Scientific Method*] *SR* **9** (1942), 547–548.

[review of P. Frank's *Between Physics and Philosophy*] *PPR* **3** (1942–3), 108–110.

"Verification, Meaning and Truth", *PPR* **4** (1943–4), 267–283.

Methodology of the Social Sciences, Oxford, 1944, pp. viii, 272 (not a translation of *Methodenlehre* 1936; reprinted New York, 1958).

[review of R. M. MacIver's *Social Causation*] *SR* **11** (1944), 117–120.

[review of A. E. Murphy's *The Use of Reason*] *SR* **11** (1944), 511–513.

[review of R. Lepley's *Verifiability of Value*] *SR* **11** (1944), 516–517.

"Concerning Mr. Nagel's Critical Comments", *PPR* **5** (1944–5), 69–74.

"A Note on Mr. Bayliss's Discussion", *PPR* **5** (1944–5), 96–97.

"Discussion of Mr. Nagel's Rejoinder", *PPR* **5** (1944–5), 351–353.

[review of W. Ebenstein's *Pure Theory of Law*] *American Political Science Review* (1945) (Oct.).

"The Nature of Scientific Method", *SR* **12** (1945), 464–480.

"Scientific Procedure and Probability", *PPR* **6** (1945–6), 47–66.

"On the Nature of Inductive Inference", *PPR* **6** (1945–6), 603–609.

Metodologia de las ciencias sociales, Mexico City, 1946 [S.T. by E. Imoz of *Methodenlehre* 1936 with a 2 pp. note at the end dated "New York, February 1946"].

[review of J. R. Baker's *Science and the Planned State*] *SR* **13** (1946), 381–382.

[review of B. Russell's *A History of Western Philosophy*] *PPR* **7** (1946–7), 461–466.

"Problems of Philosophical Education", *Freedom and Experience*, eds. S. Hook and M. R. Kovitz (Essays presented to H. M. Kallen), Ithaca and New York, 1947, pp. 221–238.

"Observations on the Ivory Tower", *SR* **14** (1947), 285–303.

"Three Meanings of Truth", *JP* **45** (1948), 337–350.

"Rudolf Carnap's Analysis of Truth", *PPR* **9** (1948–9), 295–299.

"Cassirer's Theory of Scientific Knowledge", *The Philosophy of Ernst Cassirer*, ed. P. Schilpp (The Library of Living Philosophers), Evanston Ill., 1949, pp. 185–213.
"The Issue of Ethical Neutrality in Political Science", *SR* **16** (1949), 344–352.
"Basic Issues in Logical Positivism", ed. M. Farber, Buffalo, 1950, pp. 565–588.
"John Dewey's Theory of Inquiry", *JP* **56** (1959), 826–836.
The Infinite in Mathematics, ed. B. McGuinness, with an Introduction by E. Nagel, Dordrecht and Boston, 1978 [the present volume].

BIBLIOGRAPHY OF WORKS CITED IN THE PRESENT VOLUME

Ackermann, W., 1924, 'Begründung des "tertium non datur" mittels der Hilbertschen Theorie der Widerspruchsfreiheit', *Math. Ann.* **93**, 1–36; 1928, 'Zum Hilbertschen Aufbau der reellen Zahlen', *Math. Ann.* **99**, 118–133. See also under Hilbert.

Baer, R., 1928, 'Über ein Vollständigkeitsaxiom in der Mengenlehre', *Math. Zeitschr.* **27**, 536–539.

Baldus, R., 1928, 'Zur Axiomatik der Geometrie, I. Über Hilberts Vollständigkeitsaxiom', *Math. Ann.* **100**, 321ff.

Becker, O., 1923, 'Beiträge zur phänomenologischen Begründung der Geometrie und ihrer physikalischen Anwendungen', *Jahrbuch für Philosophie und phänomenologische Forschung* **6**, 385–560; 1927, 'Mathematische Existenz. Untersuchungen zur Logik und Ontologie mathematischer Phänomene', *ibid.* **8**, 439–809f. (also available separately); 1928, 'Das Symbolische in der Mathematik', *Blätter für deutsche Philosophie* **1**, 329–348.

Behmann, H., 1922, 'Beiträge zur Algebra der Logik, insbesondere zum Entscheidungsproblem', *Math. Ann.* **86**, 163–229; 1923 and 1927, 'Entscheidungsproblem und Logik der Beziehungen', *Jahresb. d. Deutsch. Math.-Ver.* **32**, 66f., and **36**, 17f.; Sept. 1929, 'Zu den Widersprüchen der Logik und Mengenlehre', lecture to the Deutsch. Math.-Ver. in Prague, *Jahresb. d. Deutsch. Math.-Ver.* **40**, 37ff.

Bernays, P., 1922, 'Über Hilberts Gedanken zur Grundlegung der Arithmetik', *Jahresb. d. Deutsch. Math.-Ver.* **31**, 10–19; 1923, 'Erwiderung auf die Note von Herrn Aloys Müller "Zahlen als Zeichen" ', *Math. Ann.* **90**, 159–163 (cf. also *ibid.*, pp. 153–158); 1926, 'Axiomatische Untersuchung des Aussagenkalküls der *Principia Mathematica*', *Math. Zeitschr.* **25**, 305–320; 1930, 'Die Philosophie der Mathematik und die Hilbert'sche Beweistheorie', *Blätter f. deutsche Philosophie* **4**, 326–367.

Bernays, P. and Schönfinkel, M., 1928, 'Zum Entscheidungsproblem der mathematischen Logik', *Math. Ann.* **99**. 342–372.

Bernstein, F., 1905, 'Über die Reihe der transfiniten Ordnungszahlen', *Math. Ann.* **60**, 187–193; 1919, 'Die Mengenlehre Georg Cantors und der Finitismus', *Jahresber. d. Deutsch. Math.-Ver.* **28**, 63–78.

Bolzano, B., *Paradoxien des Unendlichen*, published from his literary remains by Fr. Prihonsky in 1851. New edition by A. Höfler, annotated by H. Hahn, as vol. 99 of Philosophische Bibliothek, Leipzig 1920.

Borel, E., 1898, *Leçons sur la théorie des fonctions*, Paris, 3rd edition, 1928.

Brentano, F., 'Vom ens rationis', dissertation from his literary remains, published by Oskar Kraus in vol. 193 of Philosophische Bibliothek, p. 238ff., Leipzig 1925.

Brouwer, L. E. J., 1911, 'Beweis der Invarianz der Dimensionszahl', *Math. Ann.* **70**, 161–165; 'Intuitionisme en formalisme', inaugural lecture, Groningen 1912 (English translation 1913, 'Intuitionism and Formalism', *Bull. Amer. Math. Soc.* **20**, 81–96); 1919, 'Intuitionistische Mengenlehre', *Jahresb. d. Deutsch. Math.-Ver.* **28**, 203–

208 (also in *Kon. Akad. v. Wetensch. Amsterdam, Proceedings*, **23**, (1920) 949–954); 1923, *Begründung der Funktionenlehre unabhängig vom logischen Satz vom ausgeschlossenen Dritten*. I. Teil, Amsterdam; 1925, 'Zur Begründung der intuitionistischen Mathematik' I, *Math. Ann.* **93**, 224–257; 1926, op. cit. II, *ibid*, **95**, 453–472; 1927, op. cit., III, *ibid.* **96**, 451–488; 1925, 'Intuitionistische Zerlegung mathematischer Grundbegriffe', *Jahresb. d. Deutsch. Math.-Ver.* **33**, 251–256; 1928, 'Intuitionistische Betrachtungen über der Formalismus,' *Sitzungsber. d. Preuss. Akad. d. Wissensch., Phys. math. Kl.*, pp. 48–52; 1929, 'Mathematik, Wissenschaft und Sprache', *Monatshefte f. Math. u. Phys.* **36**, 153–164.

Burali-Forti, C., 1897, 'Una questione sui numeri transfiniti', *Rendiconti del Circolo di Matematica di Palermo* **11**, 154–164.

Burkamp, W., 1927, *Begriff und Beziehung. Studien zur Grundlegung der Logik*, Leipzig.

Cantor, G., 1883, *Grundlagen einer allgemeinen Mannichfaltigskeitslehre*, Leipzig; 1887, and 1888, 'Mitteilungen zur Lehre vom Transfiniten', I and II, *Zeitschr. f. Phil. u. phil. Kritik* **91**, 81–125, 252–270 **92**, 240–265; 1895, 'Beiträge zur Begründung der transfiniten Mengenlehre', I, *Math. Ann.* **46**, 481–512; 1897, op. cit. II, *ibid.* **49**, 207–246. (1915, *Contributions to the Founding of the Theory of Transfinite Numbers*, tr. P. E. B. Jourdain, Open Court, Chicago and London.)

Carnap, P., 1922, *Der Raum. Ein Beitrag zur Wissenschaftslehre*, supplementary pamphlets to *Kant-Studien* no. 56, Berlin; 1927, 'Eigentliche und uneigentliche Begriffe', *Symposion* **1**, 355–374; 1928, *Der logische Aufbau der Welt*, Berlin-Schlachtensee (1967, *The Logical Structure of the World*, tr. R. A. George, London); 1929, *Abriss der Logistik*, Schriften zur wissenschaftlichen Weltauffassung, vol. 2, Vienna.

Cassirer, E., 1929, *Philosophie der symbolischen Formen*, 3 Teil. Phänomenologie der Erkenntnis, Berlin (1953–1955, *The Philosophy of Symbolic Forms*, tr. C. Hendel, Oxford and Yale).

Chwistek, L., 1922, 'Über die Antinomien der Prinzipien der Mathematik', *Math. Ann.* **14**, 236–243; 1923–1925, 'The Theory of Constructive Types', Principles of Logic and Mathematics, Part I and II, extracted from the *Annales de la Société Polonaise de Mathematique*, Cracow; 1926, 'Über die Hypothesen der Mengenlehre', *Math. Zeitschr.* **25**, 439–473.

Dedekind, R., 1872, *Stetigkeit und irrationale Zahlen*, 5th edition, Brunswick 1927; 1887, *Was sind und was sollen die Zahlen?*, 4th edition, Brunswick 1918. (Together in English 1901, *Essays on the Theory of Numbers*, tr. W. W. Benman, Open Court, Chicago.)

Dingler, H., 1919, 'Über die axiomatische Grundlegung der Lehre vom Ding', *Jahresber. d. Deutsch. Math.-Ver.* **28**, 138–158.

Dubislav, W., 1926, 'Über das Verhältnis der Logik zur Mathematik', *Ann. d. Philos. u. philos. Kritik* **5**, 193–208; 1929, 'Elementarer Nachweis der Widerspruchslosigkeit des Logikkalküls', *Journ. f. Math.* **101**.

Einstein, A., 1921, *Geometrie und Erfahrung*, Berlin (1922, *Sidelights on Relativity: II – Geometry and Experience*).

Finsler, P., 1926, 'Über die Grundlegung der Mengenlehre, I. Teil, Die Mengen und ihre Axiome', *Math. Zeitschr.* **25**, 683–713.

Fraenkel, A., 1912, 'Axiomatische Begründung von Hensels p-adischen Zahlen', *Journ. f. Math.* **141**, 43–76; 1927, *Zehn Vorlesungen über die Grundlegung der Mengenlehre*, Wissensch. u. Hypoth., vol. 31, Leipzig and Berlin; 1928, *Einleitung in die Mengenlehre*, 3rd edition, Berlin (1958, *Foundations of Set Theory*, North Holland).

Frege, G., 1884, *Die Grundlagen der Arithmetik*, Breslau (1953, *Foundations of Arithmetic*, tr. J. L. Austin, Oxford); 1892, 'Über Sinn und Bedeutung', *Zeitschr. f. Phil. u. phil. Kritik*. **100**, 25–50 ('On Sense and Reference', tr. M. Black, in M. Black and P. T. Geach, *Translations from the Philosophical Writings of Gottlob Frege*, 2nd edition 1960); 1893 and 1903, *Grundgesetze der Arithmetik*, I and II respectively, Jena (selections in English in M. Black and P. T. Geach, op. cit.).

Gauss, C. F., 1880, *Briefwechsel zwischen Gauss und Bessel*, Leipzig.

Geiger, M., 1921, 'Fragment über den Begriff des Unbewussten und die psychische Realität', *Jahrbuch f. Philosophie u. phänomenol. Forschung* **4**, 1–137; 1924, *Systematische Axiomatik der Euklidischen Geometrie*, Augsburg.

Gödel, K., 'Über formal unentscheidbare Sätze der Principia Mathematica und verwandter Systeme I', *Monatshefte f. Math. u. Phys.* XXXVIII. 1 ,(1931), 173–198· (1967 in van Heijenoort, J., ed., *From Frege to Gödel*).

Grelling, K., 1924, *Mengenlehre*, Math. phys. Bibl., no. 58, Leipzig and Berlin.

Hardy, G. H. A., 1903, 'A Theorem Concerning the Infinite Cardinal Numbers', *Quart. Journ. of Pure and Appl. Math.* **35**, 87–94.

Hartmann, N., 1925, *Grundzüge der Metaphysik des Erkennens*, 2nd edition, Berlin.

Hartogs, F., 1915, 'Über das Problem der Wohlordnung', *Math. Ann.* **76**, 438–443.

Hasse, H. and Scholz, H., 1928, 'Die Grundlagenkrisis der griechischen Mathematik', *Kant-Studien* **33**, 4–34.

Hausdorff, F., 1914, *Grundzüge der Mengenlehre*, 2nd edition, Leipzig 1927.

Helmholtz, H., 1895 'Zählen und Messen', *Wissensch. Abhandl.* III, 356–391.

Hilbert, D., 1899, *Grundlagen der Geometrie*, Wissensch. u. Hypoth., vol. 7, 6th edition, Leipzig and Berlin 1923 (1921, *The Foundations of Geometry*, tr. E. J. Townsend, Open Court, Chicago); 1890, 'Über die Theorie der algebraischen Formen', *Math. Ann.* **36**, 473–534; 1922, 'Neubegründung der Mathematik. Erste Mitteilung', *Abh. a. d. Math. Sem. d. Hamb. Univ.* **1**, 157–177; 1923, 'Die logischen Grundlagen der Mathematik', *Math. Ann.* **88**, 151–165; 1925, 'Über das Unendliche', *Math. Ann.* **95**, 161–190; 1928, 'Die Grundlagen der Mathematik', *Abh. a. d. Math. Sem. d. Hamb. Univ.* **6**, 65–92 (with notes by H. Weyl and an appendix by P. Bernays).

Hilbert, D. and Ackermann, W., 1928, *Grundzüge der theoretischen Logik*, Berlin (1950, *Principles of Mathematical Logic*).

Hjelmslev, J., 1923, 'Die natürliche Geometrie', *Abh. a. d. Math. Sem. d. Hamb. Univ.* **2**, 1–36.

Hölder, O., 1914, *Die Arithmetik in strenger Begründung*, 2nd edition, Berlin 1929; 1924, *Die mathematische Methode*, Berlin.

Huntington, E. V., 1924, 'A New Set of Postulates for Betweenness with Proof of Complete Independence', *Transact. Amer. Math. Soc.* **26**, 257–282.

Husserl, E., 1891, *Philosophie der Arithmetik*, Leipzig; 1900–1901, *Logische Untersuchungen, I & II*, 3rd edition, Halle 1922 (1970, *Logical Investigations*, tr. J. N. Findlay, London); 1933, *Ideen zu einer reinen Phänomenologie und phänomenologischen Philosophie*', Halle (1931, *Ideas: General Introduction to Pure Phenomen-*

ology, tr. W. R. Boyle Gibson, London and New York); 1929, 'Formale und transzendentale Logik', *Jahrb. f. Phil. u. phänomenolog. Forsch.* **10**, Halle, pp. 1–298 (1969, *Formal and transcendental Logic*, tr. D. Cairns, The Hague).

Kant, I., 1781 *Kritik der reinen Vernunft*, 1st edition (standard translation by N. Kemp Smith, 1929, *Critique of Pure Reason*, London).

Klein, F., 1890, 'Zur Nicht-Euklidischen Geometrie', *Math. Ann.* **37**, 544–572.

König, J., 1905, 'Zum Kontinuumproblem', *Math. Ann.* **60**, 177–180, 462; 1914, *Neue Grundlagen der Logik, Arithmetik und Mengenlehre*, Leipzig.

Kronecker, L., 1899, 'Über den Zahlbegriff', *Werke III*, 1, Leipzig, p. 249ff.

Leibniz, G. W. (C. I. Gerhardt, ed.), 1849, *Mathematische Schriften*, Berlin.

Lewy, P., April 1928, 'Sur le principe du tiers exclu et sur les théorèmes non susceptibles de démonstration', *Revue de Métaphysique et de Morale*, quoted in Borel, op. cit., p. 265ff.

Löwenheim, L., 1915, 'Über Möglichkeiten im Relativkalkül', *Math. Ann.* **76**, 447–470.

Menger, K., 1929, 'Zur Dimensions- und Kurventheorie', *Monatshefte f. Math. u. Phys.* **36**, 411; 1928, *Dimensionstheorie*, Berlin; 1928, 'Bemerkungen zu Grundlagenfragen', I. Über Verzweigungsmengen; and III. Über Potenzmengen, *Jahresb. d. Deutsch. Math.-Ver.* **37**, 213–226 and 303–308 respectively; 1929, review of 3rd edition of Fraenkel's *Einleitung*, in *Monatshefte f. Math. u. Phys.* **36**, Literaturberichte p. 7.

Moore, E. H., 1910, *Introduction to a Form of General Analysis*, New Haven.

Müller, A., 1923, 'Über Zahlen als Zeichen', *Math. Ann.* **90**, 153–158 (cf. also *ibid.*, p. 163).

v. Neumann, J., 1925, 'Eine Axiomatisierung der Mengenlehre', *Journ. f. Math.*, p. 219–240; 1927, 'Zur Hilbertschen Beweistheorie', *Math., Zeitschr.* **26**, 1–46; 1928, 'Die Axiomatisierung der Mengenlehre', *ibid.*, **27**, 669–752.

Neurath, O., 1909, 'Eindeutigkeit und Kommutivität des logischen Produktes "ab" ', *Arch. f. syst. Phil.* **15**, 104–106; 1910, 'Definitionsgleichheit und symbolische Gleichheit', *ibid.* **16**, 142–144.

Nicod, J. G. P., 1917–1920, 'A Reduction in the Number of Primitive Propositions of Logic', *Proc. Cambridge Philos. Soc.* **19**, 32–41.

Pasch, M., 1882, *Vorlesungen über neuere Geometrie*, Leipzig, New edition, Berlin 1926; 1917, 'Grundfragen der Geometrie', *Journ. f. Math.* **147**, 184–190.

Peano, G., 1889, *Arithmetices principia nova methodo exposita*, Turin (1973, *Selected works*, tr. and ed. H. C. Kennedy, London); 1890, 'Sur une courbe qui remplit toute une aire plane', *Math. Ann.* **36**, 157–160.

Poincaré, H., 1903, *La Science et l'Hypothèse* (1905, *Science and Hypothesis*); 1909, 'Réflexions sur les deux notes précédentes', *Acta Mathematica* **32**, 195–200; 1910, *Über transfinite Zahlen* (six lectures on selected topics from pure mathematics and mathematical physics), Leipzig and Berlin, p. 45ff.; 1908, *Science et Méthode* (1914, *Science and Method*, tr. F. Maitland, London).

Ramsey, F. P., 1927, 'The foundations of mathematics', *Proc. London Math. Soc.* (2) **25**, 338–384; 1929, 'On a Problem of Formal Logic', *ibid.* **30**, 264–286.

Russell, B., 1903, *The Principles of Mathematics*, Cambridge; 1919, *Introduction to Mathematical Philosophy*, London. See also under Whitehead.

Scholz, H., 1928, 'Warum haben die Griechen die Irrationalzahlen nicht aufgebaut?', *Kant-Studien* **33**, 35–72, See also under Hasse.

Schönfinkel, M., see under Bernays.

Schröder, E., 1873, *Lehrbuch der Arithmetik und Algebra*, Vol. 1, Leipzig.

Sheffer, H. M., 1913, 'A Set of Five Independent Postulates for Boolean Algebras, with Application to Logical Constants', *Transact. Amer. Math. Soc.* **14**, 481–488.

Sierpinski, W., 1924, 'Sur l'hypothèse du continu', *Fundamenta Mathematicae* **5**, 177–187.

Skolem, Th., 1920, 'Logisch-Kombinatorische Untersuchungen über die Erfülbarkeit oder Beweisbarkeit mathematischer Sätze nebst einem Theoreme über dichte Mengen', *Skrifter utgit av Videnskapsselskapet i Kristiania, Math.-naturw. Klasse*, no. 4, pp. 1–36; 1923, 'Einige Bemerkungen zur axiomatischen Begründung der Mengenlehre', scientific lectures given at the 5th Congress of Scandinavian mathematicians in Helsinki 1922, pp. 217–232.

Stammler, G., 1925, *Der Zahlbegriff seit Gauss*, Halle; 1928, *Begriff, Urteil, Schluss*, Halle.

Stolz, O., 1885, *Vorlesungen über allgemeine Arithmetik I Teil*, Leipzig.

Tarski, A., 1925, 'Sur les ensembles finis', *Fundamenta Mathematicae* **6**, 45–95.

Urysohn, P., 1922, 'Les multiplicités cantoriennes', *Comptes Rendus* **175**, 440ff.

Veblen, O., 1908, 'Continuous Increasing Functions of Finite and Transfinite Ordinals', *Transact. Amer. Math. Soc.* **9**, 280–292.

Voss., A., 1913, *Über das Wesen der Mathematik*, 2nd edition, Berlin.

Waismann, F., *Logik, Sprache, Philosophie*, Published in German, Stuttgart, 1976 [E.T. *The Principles of Linguistic Philosophy*, London, 1965].

Weyl, H., 1918, *Das Kontinuum. Kritische Untersuchungen über die Grundlagen der Analysis*, Leipzig; 1919, 'Der circulus vitiosus in der heutigen Begründung der Analysis', *Jahresber. d. Deutsch. Math.-Ver.* **28**, 85–92; 1921, 'Über die neue Grundlagenkrise der Mathematik', *Math. Zeitschr.* **10**, 39–79; 1924, 'Randbemerkungen zu Hauptproblemen der Mathematik', *ibid.* **20**, 131–150; 1925, 'Die heutige Erkenntnislage in der Mathematik', *Symposion* **1**, 1–32 (also separately as pamphlet 3 of *Symposion*); 1926, *Philosophie der Mathematik und Naturwissenschaft*, Munich and Berlin (1949, *Philosophy of Mathematics and Natural Science*, tr. O. Helmer, Princeton).

Whitehead, A. N. and Russell, B., 1910–1913, *Principia Mathematica*, Cambridge, 2nd edition 1925–1927.

Wittgenstein, L., 1922, *Tractatus Logico-Philosophicus*, London (previously in German only, 1921, *Ann. der Natur- und Kulturphilosophie* **14**). Later English translation [1]1961,[2] 1971.

Zermelo, E., 1904, 'Beweis, dass jede Menge wohlgeordnet werden kann', *Math. Ann.* **59**, 514–516; 1908, 'Neuer Beweis für die Wohlordnung', *ibid.* **65**, 107–128; 1908, 'Untersuchungen über die Grundlagen der Mengenlehre', I, *ibid.*, pp. 261–281.

INDEX OF NAMES

VIENNA CIRCLE COLLECTION

1. Otto Neurath, *Empiricism and Sociology*. Edited by Marie Neurath and Robert S. Cohen. With a Section of Biographical and Autobiographical Sketches. Translations by Paul Foulkes and Marie Neurath. 1973, xvi+473 pp., with illustrations. ISBN 90 277 0258 6 (cloth), ISBN 90 277 0259 4 (paper).

2. Josef Schächter, *Prolegomena to a Critical Grammar*. With a Foreword by J. F. Staal and the Introduction to the original German edition by M. Schlick. Translated by Paul Foulkes. 1973, xxi+161 pp. ISBN 90 277 0296 9 (cloth), ISBN 90 277 0301 9 (paper).

3. Ernst Mach, *Knowledge and Error. Sketches on the Psychology of Enquiry*. Translated by Paul Foulkes. 1976, xxxviii+393 pp. ISBN 90 277 0281 0 (cloth), ISBN 90 277 0282 9 (paper).

5. Ludwig Boltzmann, *Theoretical Physics and Philosophical Problems. Selected Writings*. With a Foreword by S. R. de Groot. Edited by Brian McGuinness. Translated by Paul Foulkes. 1974, xvi+280 pp. ISBN 90 277 0249 7 (cloth), ISBN 90 277 0250 0 (paper).

6. Karl Menger, *Morality, Decision, and Social Organization. Toward a Logic of Ethics*. With a Postscript to the English Edition by the Author. Based on a translation by E. van der Schalie. 1974, xvi+115 pp. ISBN 90 277 0318 3 (cloth), ISBN 90 277 0319 1 (paper).

7. Béla Juhos, *Selected Papers on Epistemology and Physics*. Edited and with an Introduction by Gerhard Frey. Translated by Paul Foulkes. 1976, xxi+350 pp. ISBN 90 277 0686 7 (cloth), ISBN 90 277 0687 5 (paper).

8. FRIEDRICH WAISMANN, *Philosophical Papers*. Edited by Brian McGuinness with an Introduction by Anthony Quinton. Translated by Hans Kaal (Chapters I, II, III, V, VI and VIII and by Arnold Burms and Philippe van Parys. 1977, xxii+190 pp. ISBN 90 277 0712 X (cloth), ISBN 90 277 0713 8 (paper).

Batch number: 10761599

Printed by Printforce, the Netherlands